"The COVID-19 pandemic left its scar on the very core of humanity. The experiences, challenges, implications, and lessons impacted the world during the pandemic. The ways in which the social structures of our societies adapted and continue to change deserve further study from the multidisciplinary writers across diverse regions. This book fills an important gap in documenting the impact of the pandemic on humanity. The book's essays chronicle cross-cultural accounts about the diversity of life experience. This book is highly recommended."
Olabisi D. Akinkugbe, Viscount Bennett Professor of Law,
Schulich School of Law, Dalhousie University, Canada

"This book highlights several singularities that this 'black swan' event has revealed as a consequence of its stress on humanity:

- Complex thinking is needed to tackle the entanglement of cause and effect within domains of knowledge to educate people about the pandemic to prevent magical thinking from taking over.
- Public health science is not the holy grail for managing a health crisis: There is an urgent need to walk the talk of collaborative ways to implement public policy.
- The post COVID-19 era must be defined by a new shared model of governance, which is bottom-up and respectful of the rights of people and communities.

Whoever reads this book will be better prepared to understand the next pandemic and contribute to its resolution as a global citizen."
Jean-Jacques Bernatas, General practitioner, public health expert,
and founder of Human Touch for Health, France

"In this fascinating survey of Covid's worldwide impacts, Nanwani and Loxley detail how all sectors of societies worked hard to adapt to this medical emergency. They show how virtual replaced in-person communication in workplaces, courtrooms and schools, and how medical facilities and responders were affected. With examples from various countries including India, Peru, Kenya, New Zealand, Philippines, and the United States, this book provides a uniquely valuable compendium of what occurred."
Paul Hirsch, Professor of Management and Organization,
Kellogg School of Management, Northwestern University, USA

"As we venture into this post-pandemic era, there is a great deal to be learned about how societies created ways of putting aside their differences to deal with the global crisis cooperatively. *Social Structure Adaptation to COVID-19: Impact on Humanity* offers insights into how nations found ways of working together to provide economic resources and develop cultural insights. The book is well worth reading and returning to many times. It provides a rich source of information for those wishing to learn from the past on how to address problems of common concern in the future."

Marie Huxtable, Visiting Research Fellow at the University of Cumbria, UK, and Editor of the *Educational Journal of Living Theories*

"This thoughtfully edited volume offers key insights into the impact of the COVID-19 pandemic on societies and individuals across the world, delving into diverse areas such as work, learning, and innovations in social media. The essays capture the extraordinary responses that emerged in the face of what the editors believe could be an outlier or a "black swan" event. The book's strength lies in its diverse range of perspectives from contributors with global expertise. Using a systems analysis approach to mapping societal changes is what sets apart *Social Structure Adaptation to COVID-19* from other books. Its comprehensive approach to understanding the pandemic's impact on society, from reflections on governance and institutions to individuals and their transformative life experiences, is worth reading."

Eva-Maria Nag, Senior Executive Editor of Global Policy Journal and Co-Director of the Global Policy Institute, School of Government and International Affairs, Durham University, UK

"This book not only expands the reader's understanding of life and work during the pandemic but also serves as a cleansing and therapeutic salve to those battered by the lockdowns and other pandemic-related restrictions and challenges."

Colin B. Picker, Professor of Law and Executive Dean, Faculty of Business and Law, University of Wollongong, Australia, and Co-Founder of the Society of International Economic Law

"With COVID-19 we lived through one of the most extensive societal disruptions ever. This book provides a comprehensive analysis of how our social systems responded. It is a treasure box of insights about the adaptation of society to humanity."

Robert E. Quinn, Professor Emeritus of Management and Organizations, Ross School of Business University of Michigan, USA

"*Social Structure Adaptation to COVID-19: Impact on Humanity* is a timely exploration of the far-reaching effects of the COVID-19 pandemic on our global society. This thought-provoking book brings together a diverse group of 27 essayists who provide unique perspectives on how social systems have adapted and responded to COVID-19. The book offers valuable insights into societies' resilience and transformative power through its comprehensive analysis of key areas such as hybrid work, virtual learning, big data, and mass communication. A must-read for scholars, policymakers, and anyone seeking a deeper understanding of the pandemic's impact on our social structures."

Swaroop Sampat Rawal, Visiting faculty, Sardar Patel University, India, and Academic Council Member, Rishihood University, India

"This book exposes the COVID-19 experience felt around the world from 2020 to 2023. COVID-19 impacted the way societies survived the pandemic, taking into account four fundamental aspects: social relations; government capacity to respond; labor force and economy; and culture. It is clear to me that the pandemic placed us in a trial-and-error mode, leaving with us the lesson that we must be agile and flexible when facing change. This book serves as a reflection point for society to make better decisions and take actions when meeting future crises."

Grace Ximena Villanueva-Paredes, Professor in the Faculty of Economic-Administrative Sciences, Universidad Católica de Santa María, Peru

"*Social Structure Adaptation to COVID-19: Impact on Humanity* is perhaps the most comprehensive book I can imagine that compiles many individual experiences and reflections from different corners of the world during the pandemic. The essayists, from different walks of life and ages, bring with them unique perspectives in sharing personal views about the pandemic, ranging from uncertainty, solitude, global citizenship, individual liberty versus collective responsibility, to the role and accountability of public institutions. This book is worth reading by anyone who wishes to know more about the societal and transformative impacts of the pandemic on everyone in the world."

Nurina Widagdo, International development practitioner, based in Indonesia, with experience working with the United Nations and major government and nongovernment organizations

"Suresh Nanwani and William Loxley provide a multidisciplinary effort by various contributors on how COVID-19 affected societal functions, and how the resulting challenges and changes will make society more complex. This book is a delightful guide of useful and important information for those of us who want to know the sociological influences of the pandemic."

Jiejin Zhu, Professor of International Studies, School of International Relations and Public Affairs, Fudan University, China

SOCIAL STRUCTURE ADAPTATION TO COVID-19

Social Structure Adaptation to COVID-19 offers global, interdisciplinary perspectives that examine how the COVID-19 pandemic has altered the development trajectory of schools, public health, the workforce, and technology adoption. It explores social themes in society, economy, policy, and culture and draws on a social framework to describe key functions of societal adaptation to the pandemic.

Edited by Suresh Nanwani and William Loxley, the volume is grounded in the study of system components and their objectives to improve overall well-being given the ill effects of the COVID-19 pandemic. Chapters explore interconnected social networks and how sectors restructured themselves to stabilize or transform society. International contributors from 20 countries offer case studies that highlight key themes including personal connectivity, societal equality, well-being, big data, and national resilience. They predict how impactful the pandemic might be in reshaping the future and assess how the COVID-19 pandemic has affected school system shutdown, public health collapse, business closures, public policy failure, and technology-driven social media acceleration.

Offering insights into how institutions and sectors work together in times of crisis, and how COVID-19 has restructured social behavior, *Social Structure Adaptation to COVID-19* will be valuable reading for scholars and students of sociology, political science, anthropology, comparative international development, psychology, and education. It will also be of interest to policymakers concerned with education, work and organizations, and media and technology.

Suresh Nanwani is Professor in Practice in the School of Government and International Affairs at Durham University, the United Kingdom. He has more than 35 years of development work experience in international organizations, including the World Bank, the Asian Development Bank, and the European Bank for Reconstruction and Development. He is a certified mediator with the Center for Effective Dispute Resolution, the United Kingdom, and an associate certified coach with the International Coaching Federation. He is the author of *Organization and Education Development: Reflecting and Transforming in a Self-Discovery Journey* (Routledge, 2021).

William Loxley is a former principal education specialist at the Asian Development Bank and executive director of the International Association for the Evaluation of Academic Achievement (IEA) headquartered in the Netherlands. He is now retired and living in the Philippines, and he maintains an advisory role in education at the above institutions. He also previously worked at the World Bank, the Ford Foundation, the Fulbright Program in Vietnam, and the Peace Corps in the Philippines.

The COVID-19 Pandemic Series
Series Editor: J. Michael Ryan

This series examines the impact of the COVID-19 pandemic on individuals, communities, countries, and the larger global society from a social scientific perspective. It represents a timely and critical advance in knowledge related to what many believe to be the greatest threat to global ways of being in more than a century. It is imperative that academics take their rightful place alongside medical professionals as the world attempts to figure out how to deal with the current global pandemic, and how society might move forward in the future. This series represents a response to that imperative.

Titles in this Series:

Inequalities, Youth, Democracy and the Pandemic
Edited by Simone Maddanu and Emanuele Toscano

COVID-19 and the Right to Health in Africa
Edited by Ebenezer Durojaye and Roopanand Mahadew

Transformations in Social Science Research Methods during the COVID-19 Pandemic
Edited by J. Michael Ryan, Valerie Visanich and Gaspar Brändle

Social Structure Adaptation to COVID-19
Impact on Humanity
Edited by Suresh Nanwani and William Loxley

For more information about this series, please visit: www.routledge.com/Routledge-Handbooks-in-Religion/book-series

SOCIAL STRUCTURE ADAPTATION TO COVID-19

Impact on Humanity

Edited by Suresh Nanwani and William Loxley

LONDON AND NEW YORK

Designed cover image: Suresh Nanwani

First published 2024
by Routledge
4 Park Square, Milton Park, Abingdon, Oxon OX14 4RN

and by Routledge
605 Third Avenue, New York, NY 10158

Routledge is an imprint of the Taylor & Francis Group, an informa business

© 2024 selection and editorial matter, Suresh Nanwani and William Loxley; individual chapters, the contributors

The right of Suresh Nanwani and William Loxley to be identified as the authors of the editorial material, and of the authors for their individual chapters, has been asserted in accordance with sections 77 and 78 of the Copyright, Designs and Patents Act 1988.

The Open Access version of this book, available at www.taylorfrancis.com, has been made available under a Creative Commons Attribution-Non Commercial-No Derivatives (CC-BY-NC-ND) 4.0 international license.

Trademark notice: Product or corporate names may be trademarks or registered trademarks, and are used only for identification and explanation without intent to infringe.

British Library Cataloguing-in-Publication Data
A catalogue record for this book is available from the British Library

ISBN: 978-1-032-69025-4 (hbk)
ISBN: 978-1-032-69026-1 (pbk)
ISBN: 978-1-032-69027-8 (ebk)

DOI: 10.4324/9781032690278

Typeset in Gaillard
by Apex CoVantage, LLC

For all of us, our loved ones, and future generations

CONTENTS

List of figures xv
List of tables xvii
List of abbreviations xviii
List of contributors xix
Preface xxiv
    Suresh Nanwani and William Loxley
Acknowledgments xxvi
Foreword xxvii
    Colin B. Picker
Series editor foreword xxx
    J. Michael Ryan

PART I
Introduction   1

1 Society and COVID-19   3
   William Loxley and Suresh Nanwani

2 Identifying international essays about COVID-19   32
   Suresh Nanwani and William Loxley

3 Awakening from a paradise lost: experiences and lessons of the pandemic   44
   Owen Bethel

PART II
Educational opportunity and social mobility — 55

4 Introduction to the social sector — 57
William Loxley and Suresh Nanwani

5 Personal development in a time of crisis — 61
Frédéric Ysewijn

6 Tails we go, heads we stay — 65
Ariel Segal

7 Adapting to virtual education during the COVID-19 pandemic in Peru — 72
Victor Saco

8 The plight of virtual education in India due to COVID-19 — 78
Anwesha Pal

9 COVID-19 and tertiary education: experiences in Lesotho institutions — 87
Tsotang Tsietsi

10 Lifelong learning as a powerful force in the post-pandemic world — 95
William Loxley

PART III
Public policy and risk management — 103

11 Introduction to the political sector — 105
Suresh Nanwani and William Loxley

12 New Zealand: global connectivity and digital diplomacy — 108
Tracey Epps

13 Brazil in crisis mode: institutions in times of uncertainty — 115
José Guilherme Moreno Caiado

14 Legal practice in Kenya: embracing automation and e-judiciary — 122
Leyla Ahmed

15 The pandemic and post-pandemic aftershocks: whither legal education? *Shouvik Kumar Guha*	132
16 Change and continuity: COVID-19 and the Philippine legal system *Antonio G.M. La Viña*	140
17 Digital technology: a best friend for implementing COVID-19 policy in China *Li Xudong*	149
18 The goldfish and the net(work) *Nicole Mazurek*	157

PART IV
Diversity in workforce behavior — **167**

19 Introduction to the economic sector *William Loxley and Suresh Nanwani*	169
20 Entering the workforce in the COVID-19 era *S.R. Westvik*	172
21 How a Gen Y became a Gen Z at heart *Bina Patel*	179
22 Unlocking from lockdown: reframing the future through appreciative dialogue *Keith Storace*	188
23 COVID-19 and moving to the new normal *Victoria Márquez-Mees*	201
24 Financial literacy: its relevance in the education curriculum *Marie-Louise Fehun Aren*	208
25 Self-coaching for pandemic survivors *Vikram Kapoor*	219

PART V
Technology and culture · 225

26 Introduction to the cultural sector · 227
Suresh Nanwani and William Loxley

27 Will technology replace or recreate humans? · 230
Arthur Luna

28 Privacy issues in online education technologies in China · 236
Li Mengxuan

29 Digital technology during COVID-19 in global living educational theory research · 243
Jack Whitehead

30 I'm gonna let it shine: local musicians in the Virginia countryside during the pandemic · 251
Lori Udall

31 Emotional and physical isolation in a Latino community · 256
Alfred Anduze

32 Braving COVID-19 through the Gross National Happiness way in Bhutan · 264
Tshering Cigay Dorji

33 A resurrection: human connections and beyond · 272
Suresh Nanwani

PART VI
Conclusion · 281

34 Social structure adaptability to the pandemic · 283
Suresh Nanwani and William Loxley

Index · *298*

FIGURES

1.1	The many faces of change altered by COVID-19	10
1.2	Time spent on activities (outer circle 2020, inner circle 2019)	14
1.3	Covid-related central themes emerging from social, political, economic, and cultural sectors	26
2.1	Map of the world by regions	34
8.1	Rachna at her school (Ghoshpara Nishchinda Balia Vidyapith, government-sponsored)	80
8.2	Ria at her primary school in the village	80
17.1	Screenshot of Li Xudong's Health QR Code with English explanation	150
17.2	A kid in Superman costume lining up for mass testing in a residential area, China	153
22.1	Appreciative Dialogue Framework	192
22.2	Appreciative Dialogue Sequence	194
24.1	Personal finance components	214
29.1	American Education Research Association (AERA) 2022 Symposium with participants Jackie Delong, Jack Whitehead, Swaroop Rawal, Michelle Vaughan, and Parbati Dhungana (clockwise from top) on April 22, 2022	246
29.2	Participants at the 2021 conference of the American Educational Research Association with participants Jackie Delong, Jack Whitehead, Shivani Mishra, Michelle Vaughan, and Parbati Dhungana (clockwise from top)	247
30.1	Mark Maggiolo at home	252
30.2	Michelle and Rich Coon performing	253
30.3	Jam at Orlean Market	254

32.1 A *Desuup* volunteer feeds the dogs in the central district of
 Bumthang during the lockdown on January 18, 2022   267
32.2 IT teams busy at work on the night of March 19, 2020,
 Bhutan   270

TABLES

1.1	Structural elements and key sectors that promote change	21
1.2	Four key issues in national sectors arising from the pandemic	24
2.1	Writers from the globe – region and number	38
2.2	Schedule for the book project	40
2.3	Matrix for rating the pandemic on issues and themes	42
22.1	Gender, age, activity status, and preexisting mental health condition	195
22.2	Comparisons of anxiety levels pre- and post-ApDi therapy based on average anxiety scores	196
22.3	Comparisons of mindset levels pre- and post-ApDi therapy based on average mindset scores	196
22.4	Comparisons of average scores for each resilience theme pre- and post-ApDi therapy	197
34.1	Number of contributors rating COVID-19 on themes	290
34.2	Sector-on-sector support to overcome COVID-19	292

ABBREVIATIONS

ADR	alternative dispute resolution
AI	appreciative inquiry
AL	appreciative living
GNP	gross national product
GNH	gross national happiness
ICT	information and communications technology
IT	information technology
IMF	International Monetary Fund
LDCs	least developed countries
LMS	learning management system
NGO	non-governmental organization
OECD	Organization for Economic Cooperation and Development
UN	United Nations
VPN	virtual private network
WHO	World Health Organization
WTO	World Trade Organization

CONTRIBUTORS

Leyla Ahmed, a Kenyan, is a senior associate of a top law firm in Nairobi and an arbitrator. She is a fellow of the Chartered Institute of Arbitration, who handles constitutional and human rights matters and disputes in the telecommunication sectors. She has an LLM in International Trade and Investment Law from the University of Pretoria. Her chapter details how the Kenyan court system collapsed under COVID-19 restrictions and had to be quickly cobbled together in clever ways to continue processing cases.

Alfred Anduze is Puerto Rican and a medical doctor. He has written textbooks and articles focusing on quality of life, including *Social connections and your health*. His chapter looks at how Puerto Rican culture of "hugs and closeness" got sidelined as social isolation of the pandemic prevailed, but ultimately resurfaced as a strong cultural trait with the help of technology.

Marie-Louise Fehun Aren is a Nigerian citizen and a doctoral candidate at the University of Pretoria, South Africa. She has worked on numerous law and policy reform projects in Nigeria and served in the Nigerian Justice Policy Development Team. Her chapter calls for rethinking how financial literacy skills can be incorporated into school curricula, since COVID-19 has caused economic shutdowns that prevented the government from providing adequate social welfare benefits.

Owen Bethel is a citizen of the Bahamas and president and CEO of a solar renewable energy provider. He had a successful financial service sector career at IMF and World Bank and was on the Board of Directors at UNESCO. His chapter examines how COVID-19 disrupted all sectors of his island nation.

xx Contributors

José Guilherme Moreno Caiado is a Brazilian working in Germany as an adjunct professor for international trade law at the Rhine-Waal University of Applied Sciences. He is a cofounder of the PEPA/SIEL conferences that help young professionals present their ideas and help network with mentors. His chapter looks at how the Brazilian public sector dealt with the pandemic. He also discusses the linkages maintained with multilaterals during stressful times.

Tshering Cigay Dorji is a Bhutanese technologist with a PhD in computer engineering from Japan, who helped protect the livelihood of people during the pandemic. His chapter talks about the use of technology in Bhutan's fight against the pandemic through welfare programs.

Tracey Epps is a New Zealander. She is a lawyer and trade policy expert advising on WTO and free trade agreements. She teaches at the University of Otago in Dunedin, New Zealand. Her chapter reveals how virtual communication, especially via Zoom, managed to keep diplomacy functioning under the severe lockdown in New Zealand. She also talks about the extreme difficulties of international travel and communication during the pandemic.

Shouvik Kumar Guha is an Indian national working as an assistant professor at the West Bengal National University of Juridical Science, and director of the Centre for Law, Literature, and Popular Culture. He is currently working on improving legal education through the use of artificial intelligence. His chapter addresses some of the problems faced in higher education under virtual learning through the eyes of students.

Vikram Kapoor is a US citizen, a graduate of Georgetown University, president of a coaching company, and author of a popular self-coaching book. His previous experience includes coaching conflict resolution programs at the United Nations in 40 countries. His chapter looks at how self-coaching can help people survive the pandemic.

William Loxley is a US citizen working in education throughout Asia for decades as a researcher at the World Bank, principal educator in the Asian Development Bank, and executive director at the IEA in the Netherlands. His chapter looks at the span of lifelong learning as it affects not only youth and midlife decisions but especially seniors and the elderly who were disproportionately affected by the pandemic in many ways.

Arthur Luna is a Filipino citizen. He is a technology specialist working with Banco de Oro Unibank Inc. (BDO), the largest bank in the Philippines. A graduate in electronics and communications engineering, he troubleshoots

IT problems for companies in areas of communication networking and cloud computing in the Philippines and overseas. His chapter looks at whether technology will replace or recreate humans.

Victoria Márquez-Mees is a citizen of Mexico living in London, the United Kingdom. She is chief accountability officer of the European Bank for Reconstruction and Development, and professor in practice of the Latin American and Caribbean Centre at the London School of Economics. Her chapter reveals the issues faced by management in large public organizations in dealing with staff assignments throughout the COVID-19 pandemic.

Nicole Mazurek, an Australian, is a practicing lawyer in New South Wales, Australia, and England and Wales. She has worked in Russia and Poland, advising local and international financial intuitions. Her chapter looks into how government policy decisions on social distancing and general rules for the public safety could have been better handled through greater participation of stakeholders and better explanation of what actually occurred.

Li Mengxuan is a citizen of China. She is a young professional studying for her advanced law degree in the United States. She has research interests in public international law and politics. Her chapter looks at the dilemma of collecting data on students through biodata streaming in online education and using it to improve performance without giving it up to commercial interests.

Suresh Nanwani is a Singapore citizen. He is an educator, author, and writer. He has a PhD in organization development from the Philippines. He is professor in practice at Durham University, the United Kingdom. His chapter looks into human connections through appreciative inquiry and *ikigai* – the pleasure and meaning of life derived from a mindset that provides hope and deliverance.

Anwesha Pal is an Indian citizen teaching at the WB National University of Juridical Sciences in Kolkata. She ran distance education programs in rural areas during COVID-19 lockdown and experienced firsthand the learning hardships in poor communities. Her chapter chronicles ways the rural poor pursue studies in the absence of computers and ICT connections.

Bina Patel is an American who received her doctorate in conflict resolution and peacekeeping from Nova Southeastern University in the Fort Lauderdale's academic and research hub. As an organizational health strategist specializing in change management, she consults for both government and private companies. Her chapter looks into the effects of COVID-19 on attitudes and behaviors of the various generations to working conditions.

Victor Saco is a professor of International Economic Law at the Pontificia Universidad Católica del Perú (PUCP). He heads the Master's degree program on international economic law. He is a Peruvian national, a descendent of indigenous Peruvian, Chinese, and Spanish ancestry. His chapter reveals the logistical problems in higher education caused by COVID-19.

Ariel Segal is an Israeli young professional in international relations and now a project manager at a large global company focusing on sustainable development. Her chapter focuses on how the international lockdown on travel affected her ability to maintain personal contacts, complete her study, and work virtually.

Keith Storace, an Australian, is a registered psychologist with the Psychology Board of Australia, specializing in psychotherapeutic treatment in the quality of life through appreciative inquiry. He is an editorial board member for the *International Journal of Appreciative Inquiry*. His chapter looks into reframing the future through appreciative dialogue.

Tsotang Tsietsi is a citizen of Lesotho with a doctorate from the University of Cape Town, South Africa. She is a senior lecturer and coordinator of postgraduate programs in Lesotho who dealt with problems faced by first-year students, those in practical disciplines, and postgraduate research students during school closures. Her chapter details how she overcame problems faced by teachers and students as the university moved to virtual learning.

Lori Udall is an American with 30 years of experience in international development specializing in impacts on governance and finance on human development. Her chapter looks at the lives of several musicians during COVID-19 and how their artistry was affected by internet platforms and technology. She interviews musicians to see how the process of creating and playing music was reshaped by the COVID-19 experience.

Antonio G.M. La Viña is a Filipino lawyer, educator, and human rights writer; former dean of the Ateneo de Manila University School of Government; and former environmental undersecretary of the Philippines. He is the founding president of the Movement Against Disinformation and a member of the Board of Directors of Rappler Inc. His chapter studies how public sector policy influenced handling the pandemic in the country and makes comparisons of success and failure to other countries.

S.R. Westvik (Sarah Westwik) is a Norwegian young professional completing her MA in International War Studies at University College Dublin and the

University of Potsdam. Her chapter employs a magazine-style format to tell how COVID-19 university experiences affected career choices, including the digitalization of home and work social isolation that later influenced diversified opportunities in the workplace.

Jack Whitehead, a UK national, is visiting professor at the University of Cumbria and former president of the British Educational Research Association. His chapter examines how virtual communication across nations promotes living educational theory research, especially between the East and West.

Li Xudong is a citizen of China. A young professional, he is currently studying for his Master of Law degree at Northwestern University in the United States, where he focuses on corporate law. After graduation, he plans to take the New York Bar Exam. His chapter looks at how Chinese policymakers monitored COVID-19 using big data techniques of mapping locations, health status, and travel destinations of citizens.

Frédéric Ysewijn is a young professional Belgian national with a master's degree from the London School of Economics and is currently writing a book on the European Union political economy. His chapter looks at how the pandemic affected continuing education.

PREFACE

During the onslaught of COVID-19 in 2020, we coedited *Covid-19 in the Philippines: Personal Stories* (Amazon Kindle, 2021). The book contains 40 stories of residents in the Philippines, local and foreign, on their immediate experiences during the first year of the pandemic.

In 2022, we decided on a different approach in assessing the impacts of the pandemic: We identified, commissioned, and surveyed 27 professionals from various fields of expertise all over the world, to share how COVID-19 has affected their jobs, communities, and social relations. Fields of expertise included education, psychology, law, medicine, sociology, coaching, information technology, and international development.

These essays reinforce global news reports on how the pandemic unfolded by looking at personal social interaction across areas of expertise covering social, political, economic, and cultural spheres. In addition to writing essays, the contributors were surveyed on key aspects of the pandemic gleaned from the 27 essays including social networking, health, aging, and data analytics.

Survey findings complement the positions espoused in the essays, social theory, and news media that support the idea that societies come together during crises. These assessments show that in times of crisis, sectors and institutions eventually step up to serve their constituents. Governments fund and procure vaccines. Business supports public policy as do schools and families, as they all strive to maintain resilience. Many other issues facing society include environment, social class inequities, gender imbalance, and ethnic diversity that fold into these larger themes. The survey results give direction to framing hypotheses about how "black swan" events can be leveraged to improve society.

To promote COVID-19 research, the authors suggest employing systems analysis to understand how societal sectors strengthen the public good. The authors hope that readers will gain a better understanding of connections among key social processes that influence social mobility, career, public policy making, and participation in the techno-culture of the metaverse. It is likely that COVID-19 will be viewed as a major impetus in altering these societal functions of education, health, work, and social media. Institutional change in combination with demographics and environmental changes will affect diversity in social mobility, which will alter values and information flow, education, occupation, and income distribution.

This multidisciplinary book helps explain how social structure adapts to crises needed to maintain national resilience. The systems analysis approach recognizes the importance of structural change in core functions of society. A structural approach to societal organization grounded in system analysis is useful in real-world adaptation to COVID-19. The social systems approach to crisis management suggests that all parts of society together provide a powerful means to meet the demands of development in the modern world.

Suresh Nanwani and William Loxley

ACKNOWLEDGMENTS

We are grateful to everyone who contributed directly or indirectly to the writing of this book and have supported us in many ways, including the contributors who selflessly shared their thoughts, experiences, insights, and time.

We are also grateful to Joshene Bersales for helping us prepare this book.

FOREWORD

The pandemic of the early 2020s will be studied for many decades to come. Its deep and global impact, while not as catastrophic, certainly was reminiscent of the effects and reach of the prior world wars. This book is one of the first to be published in the "post-pandemic" period (though a better way to describe today is the "over-pandemic" period – during which Covid is still present and causing harm, but where now the governments and people are "over" it and simply wish to move on). As such, the book will help to slake the current thirst for understanding and insights into those grim years we all just lived through, knowing it will take time to comprehend the lessons to be adopted.

Furthermore, scientific, geopolitical, health, legal, and other macro issues of the pandemic will benefit from the inevitable deep analysis to be carried out over future decades. Those experiences and views of individuals living through the pandemic are best captured as soon as possible during and after the event. These early insights and analyses deeply shaded by individual involvement have a historical role to play over the coming years. As these initial impressions lay the groundwork for further studies, they will come to support or refute those incipient findings or wisdom.

The many unforeseen thoughts and the emotional color that bleeds through, such as those captured in this important book, become critical data for future researchers. That is why this work is such an important contribution to our understandings of those tough and challenging Covid years – both for today and for the future.

For readers today, in the immediate "post-pandemic" period, this publication plays a paramount role as we gradually rebuild lives and careers in the wake of the pandemic's destructive upheaval. Indeed, some of the main challenges of the pandemic were (i) the ubiquitous isolation caused through the repeated and long lockdowns; (ii) the social distancing rule; and (iii) the remote learning and working practices that were developed and promulgated across societies in efforts to stem the spread of Covid. When individuals and groups endure such isolation, the world mostly shrinks to one's own experiences.

To the extent we learned of others' encounters, those were themselves often remotely shared or communicated by colleagues, friends, and family or even superficially propagated by the media. In other words, the experience of the pandemic became highly individualized, while at the same time the encounters of others were but dimly and remotely felt and hence little or wrongly understood. These two experiential aspects of the pandemic no doubt have led to erroneous impressions and insights. This book's numerous individual essays by 27 essayists across the globe – from the Pacific across Asia, Middle East, Europe, Africa to the Americas – with their individual observations help to correct the lack of the direct interaction with others' lives and works during the pandemic.

Even as the title suggests the many diverse ordeals and consequences encountered during the pandemic, it is likely that many of the book's narratives and insights will strike a similar chord with the reader. In many instances, they will be ones also experienced by the reader. Sympathetic points of view are often valuable, for just as the book can share and correct erroneous understandings of others' circumstances, the stories can also confirm or legitimate individual accounts as "normal" throughout the world.

This book not only expands the reader's understanding of life and work during the pandemic but also serves as a cleansing and therapeutic salve to those battered by lockdowns and other pandemic-related restrictions and challenges. Those who felt that their challenges and hurt were unique to them or who reflected some deficiencies on their part gain a sense they were not alone. The experiences, shared by the book's essays, show the reader that the challenges and hurt were not a negative reflection on them but rather a shared universal encounter.

Finally, this foreword must also note just how complementary and suitable are the book's editors, Suresh Nanwani and William Loxley, as they balance each other's expertise, which worked well to nurture the many contributors to this study. Indeed, the book's union of these two scholars from such different fields has worked to ensure the book has broad appeal and applicability – to the lay reader and researcher alike.

Specifically, Suresh Nanwani's expertise in international development work and organization development studies and models meant he was able to draw on those decades' experiences to understand how the many different contexts fit together in the book. Complementing those broad contextual understandings, William Loxley brings out his data analytic and social development skills. This provides deep-level sociological analysis alongside the individual contributor's accounts, making it all fit together so well. It must also be thought that these two editors, alongside the many contributors, will have found the entire exercise of putting this work together to be cathartic – articulating and sharing their explorations, inquiries, and experiences each had endured through the pandemic.

Colin B. Picker
Professor of Law and Executive Dean, Faculty of Business and Law,
University of Wollongong, Australia and Co-founder, Society of
International Economic Law
June 2023

SERIES EDITOR FOREWORD

The SARS-CoV-2 virus unleashed one of the most catastrophic global health crises in living memory. The COVID-19 pandemic did the same to societies, businesses, education, relationships, and nearly every aspect of individual lives. Medical professionals have been necessary to battle this (still ongoing) crisis from a health standpoint, while social scientists and other expert professionals have been necessary to battle this (still ongoing) crisis from a social and cultural perspective. This crisis has reinforced the importance of those critical meetings points where health and society intersect.

During the early years of the COVID-19 pandemic, nearly everyone felt overwhelmed, including by daily reports of case counts, death tolls, and, later, vaccination rates. This kind of quantitative data was useful in processing large amounts of information but often missed the more personalized human touch. This brilliantly curated collection helps bring that touch by providing the reader with insights into the diversity, and similarity, of experiences from around the world.

With contributions from more than two dozen essayists, and spanning an impressive six continents, this edited collection represents one of the most truly global collections of pandemic scholarship available today. And a global collection is exactly what is needed to respond to a global crisis. Most valuable is that while each essay provides unique observations voiced from a particular location, each also provides general insights, bridging the gap between the individual reader and the global narratives presented. While COVID-19 might have made us all feel a little bit more connected in a shared global crisis, the essayists in this volume help to make us all feel quite a bit more connected with a shared global hope.

J. Michael Ryan
Series Editor, *The COVID-19 Pandemic Series*
October 2023

PART I
Introduction

1
SOCIETY AND COVID-19

William Loxley and Suresh Nanwani

Introduction

Many books will undoubtedly be written on the impacts of the COVID-19 pandemic on society. Why should this book attempt to explain COVID-19 connections to change? The answer is simple: because COVID-19 has forced a critical rethink on how to perpetuate global culture in light of existential threats to the future. Social isolation caused by COVID-19 has reduced access to new approaches in health, education, work, and human development.

Will schools adopt better learning models? Will governments stress participation and transparency? Will businesses rethink the way work is organized? Will social networks strive for inclusivity? Will cultures create new forms of human expression and awareness? Many more questions abound. To address them, the editors have curated essays by professionals from multidisciplinary areas such as education, public policy, work and societal relationships, medicine, organization development, information technology, and cultural relations.

COVID-19 may turn out to be a "black swan" event, defined as an unforeseen outlier event with severe consequences that ultimately subside. These out-of-ordinary events occur every several generations, impacting the very fabric of nations and forcing experiential issues to the forefront in every society. They allow bending the arc of progress in new directions between men and women, work and family, school and career, government and citizen, and green and brown planets. As nations endure pandemics, they get transformed into something good and bad that need to be chronicled for posterity. These game-changing events set in motion new ways societies and institutions adapt and build strategies to cope (Ryan, 2023). People alter collective values and behaviors that influence human development. This book assesses the COVID-19 experience on the world from 2020 to 2022 as it impacted daily life for all, including workers (see the book cover photo).

Recent black swan events that threaten humanity include climate and environment; fairness and equity across sub-populations; changes in ideologies that define the individual and the state; and control of digital communication technology that alters personal relationships. These events also affect the way societies progress. This is why evaluation is required to contrast the severity of COVID-19 on future trends, given the increasing problems facing the world and humanity.

The COVID-19 impact on the world can be contrasted with past and recent events, such as the HIV-AIDS crisis; the financial crisis of 2008; the wars in the Middle East and Ukraine; and the massive illegal immigration of the disadvantaged from West Africa, Central America, Middle East, and Myanmar. Nations should assess the COVID-19 pandemic relative to these slower-moving disasters.

It may turn out that the COVID-19 pandemic will not be the worst in our history. We have had world wars (World War I and World War II) in the 1900s; decimation of civilizations such as the Incas and the Mayans; plagues in ancient Egypt; and river floods in ancient China that heralded new dynasties. However, the COVID-19 pandemic was a significant event that affected everyone, as social isolation has the potential to unbalance the modern global economy. No country was spared. The COVID-19 calamity jarred the world into action on climate-environment, employer-employee relations, school and lifelong learning, public health, and the realization that technical platforms can spread misinformation on the internet and harm individuals and the common good.

It may also happen that the COVID-19 virus will remain with us through 2022 and beyond, though not as severe as when it started. It may also shift from a pandemic to an epidemic, as suggested by World Health Organization (2022). There could be new pandemics, such as mpox, as the virus is being found for the first time in people with no clear connection to its origin in Western and Central Africa. As this is the largest outbreak outside of Africa in 50 years – with confirmed cases spreading to Europe, the Americas, and Australia – it has taken scientists by surprise. In this case, however, there are vaccines already available.

In this book, we investigate some major impacts of COVID-19 – direct and indirect – on global development (Turner, 2022). How did the pandemic permanently alter the dispositions of almost eight billion people from 2020 to 2022? What are its likely long-term effects on humanity? While the book relies greatly on the professional opinions of authors and essayists, the editors also gathered information for more than two years from many sources, including the British Broadcasting Corporation (BBC), Cable National Broadcasting

Company (CNBC), Cable News Network (CNN), Channel News Asia (CNA), Deutsche Wella (DW), Al Jazeera, and Bloomberg, along with other media references such as IMF, ILO, and OECD newsletters, McKinsey & Company, and VOX EU data analytic reporting services and search engines.[1]

Chapter 1 begins by describing global development just prior to COVID-19 in early 2020. The historical context carries on through 2020, 2021, and 2022 by highlighting key events in technology, vaccine technology, quarantines, and remote learning resulting from the pandemic. Drawing on a sociological framework, the book describes the way societies provide economic resources and cultural insight to advance the common good. Readers can then examine ways pandemics have changed global society and identify major influences that predict lasting effects of COVID-19. From analyses, four themes are identified corresponding to skill learning, data-based policymaking, work and career, and educational technologies improving self-awareness. Essays from around the world are included to expound on COVID-19 issues, leading to observations on how the pandemic has changed and is changing the world.

To understand change even without cataclysmic events, we look at how society provides its own framework to ensure well-being. We also identify the pandemic's impact over several years of quarantines, given the money spent, population deaths, and vaccinations delivered (Knight & Bleckner, 2023). Finally, we look at how COVID-19 shapes behavior and explore major avenues where society might change forever as a result of COVID-19. Although the pandemic did not directly cause all the world's problems, it exacerbated them in nuanced ways. This is what the book explores.

The unsuspecting world meets the pandemic head-on

In January 2020, air travel was immensely popular, given the holidays and year-end business trips. A colleague was returning to Asia from Europe with his college daughter who planned to spend a week with him before classes resumed. He noted the unease among fellow passengers of all nationalities. Everyone was unsure about traveling, given stories of an infectious disease that easily transmits to humans. His decision to bring his daughter with him for a brief vacation was indeed a wise move. As that vacation turned into months of lockdown, he was thankful his daughter remained with him rather than be isolated in a boarding school in Europe.

Parents were also challenged when their young children asked them why they were stuck indoors, not allowed to go to school or play with their friends, and could only view the world outside through windows.[2] These two narratives during the incipient Covid days demonstrate a human quality surfacing to the fore: connection between parents and children, with values of bonding, protection, safety, resilience, and grit. These values contribute to ensure that the family stays one to survive and, in the process, nourish humanity.

Along with the coming of COVID-19 travel bans on world travel, the sudden occurrence of large infections, deaths, eventual vaccinations, large social costs, and disruptions to everyday activities would be experienced everywhere. According to WHO statistics, the COVID-19 pandemic will result in over 6.5 million deaths (with actual numbers likely three times higher), 650 million infections, billions partly vaccinated, and 17 trillion dollars spent on public health and cash payments.

IMF claims that the world GDP declined from 4.5 to 3.5 percent in 2020. ILO reports that unemployment surged dramatically, as hundreds of millions were furloughed. UN announced that extreme poverty rates increased everywhere, as did poor health, depression, obesity, and suicide.

The lost income and earnings, lost loved ones, and lost opportunities of eight billion people of all ages by geography, gender, ethnicity, and social class would lead to untold grief. Some have argued the COVID-19 statistics did not change society's trajectory much, and as such, COVID-19 could have been downplayed. However, this "survival of the fittest mentality" translated to vaccination hesitancy and refusal of social distancing. Refusing to wear face masks certainly reduced collective support to fight COVID-19 and prolonged the pandemic.

Still, there is no doubt that world development has slowed down due to COVID-19. The final cost of the lasting effects of "long covid," perhaps lasting years along with complementary health care requirements, remains unanswered. Researchers now know that between 2019 and 2021, student learning has decreased by eight months on average worldwide, and hundreds of millions of adults lost job security and income, which weakened family resilience (Bryant et al., 2022). They also now know that a few nations, such as New Zealand, Singapore, and the Nordic countries, fared reasonably well by valuing cooperation, collaboration, teamwork, and partnership in delivering care and food sharing with citizens and residents. Such collective behavior allowed the following issues to be addressed quickly:

- regain lost productivity in the business community with office and factory closures, supply chain disruptions, worker layoffs, and lost earnings;
- make up student learning deficits to raise future personal skills and opportunities;
- increase government assistance to control future pandemics by providing accurate information and mobilizing science, public opinion, and financial resources to be promoted especially in public health and social welfare; and
- adapt communication technology to inform the public during the pandemic, which permits individuals to reemerge from the quarantine to regain their sense of trust and self-confidence to pursue income, education, career, and to meet lifelong ambitions that give happiness.

The ancient Babylonians, Chinese, Egyptians, and Greeks explained war, pestilence, famine, and earthquakes as disasters caused by the displeasure of the gods of war, agriculture, and nature. In modern times, weather watchers often suggest Mother Nature's displeasure when floods, droughts, and earthquakes arise. Then, too, policymakers interpret these out-of-ordinary events as part of the long road of history that validates progress.

The *yin and yang* mentality balances the good with the bad, positive with negative, and appreciative inquiry with living. It has often resulted in a philosophy of "two steps forward and one step back" over the past 700 years, from the Black Plague in the 14th century. Consequently, numerous theories have been used to give meaning to black swan events in social, political, economic, and cultural contexts. Whatever the theory, these negative happenings are here to stay, coming every several generations and arriving even more frequently when human error is widespread.

Historical experience with disasters shows that populations are more successful in moving forward when all segments of society are aligned and fall neatly into place to reinforce one another. Nations succeed when (i) layers of social stratification are open to mobility especially through education; (ii) government leaders plan well, implement, and monitor; (iii) economies spread the wealth of nations; and (iv) culture gives meaning to major events in ways that allay fears often through efficient modern innovations. Through cooperation, pandemic-sized disasters can be better controlled, despite the god of incompetence Koalemos' efforts to increase risk.

The global setting in late 2019 and the epidemiology of COVID-19

The social, political, economic, and cultural global system was already shifting by the first decade of the 21st century (World Bank, 2021). This is expected in modern societies seeking transformative change over the long run. However, the pace of change was uneven. In the arena of science and technology, changes were coming fast and furious. In other areas, policymakers were unable to persuade society of the importance of fairness, equity, and cooperation. This was especially true in areas of social efficiency, immigration, climate change, and environmental protection. Difficulties in solving problems prevented policymakers from reinforcing common good practices, which led to a lack of confidence in government and insecurity in economics (or economic activities) across nations.

Globalization in the first two decades of the 21st century found GDP modestly growing in advanced societies and much more in developing economies, as they later expanded faster from a lower base (Kharas et al., 2020). Economists endlessly debate whether rich and poor economies eventually converge or diverge over time in economic development. The answer is not so straightforward and has major consequences for how well societies can meet the needs of the global community in a fair and sustainable manner.

So far, the provisional answer, as suggested by macro-economists like Paul Romer and Robert Lucas, says that under the right conditions, all nations can converge to similar levels along their growth trajectories. Yet in order for all nations to end up at the same level, they would have to be on the same initial growth curve. One naturally asks if the effects of COVID-19 have influenced this process by accelerating or retarding the ranking of rich and poor nations.

Over the past 20 years, population growth has been decelerating, although geographical or regional differences remained massive. Alternatively, the wealth gap was increasing rapidly. There were many small ethnic and tribal conflicts, especially in Central America, Africa, the Middle East, and South Asia. Major equity issues arising from inequality have implications for climate change on environments where poor communities, for example, have fewer options to support environmental protection. In poor nations, social services for schooling, health, public transport, and digital divide were underfunded, which only accelerates inequality. Education has become a major avenue in modernizing societies for generations to transition to varied alternatives where mobility supports change.

Senior citizens who require medical attention more than other age groups were already slowly being segregated based on their ability to pay for quality services. Meanwhile, the youth feared falling behind in learning relevant skills for employment that lead to productive fulfillment. Young people were left without school routines needed to learn lessons and gain social etiquette. Likewise, many middle-aged individuals facing midlife crises were negatively affected psychologically by COVID-19 in pursuing their careers.

Political inaction in solving key social problems like the environment, social mobility, public health quality, pay gaps, and internet safety has prevented planning and digital monitoring of the environment and social services. Science and technology were improving productivity in technical innovation, enterprise, human capital, and capital expenditure, but profits often went to winner entities that have comparative advantage. Among all this turmoil, COVID-19 entered the mélange.

In 2023, the health of populations in the advanced world (1.5 billion people) appeared adequate except for the uninsured; lacking for citizens in low- and lower-middle-income countries (3.5 billion); and abysmal for most emerging societies (3 billion). The global health sector did not face major epidemics during the first two decades of the century. However, the Ebola outbreak in West Africa in the second decade of the 21st century did present problems given the severity of the disease and speed of spread in unhygienic situations. Ebola was contained in parts of West Africa and, although very contagious and toxic, was kept isolated from most people while experimental vaccines helped control the disease. The same was true for polio and malaria.

Although the Bird Flu and Swine Flu pandemics spread intermittently around the world, they were fortunately short-lived as WHO took stern measures early on to track and shut down contagion. Avian Flu started in 1997 and ran for 10 years as bird migration spread the disease. The Swine Flu pandemic started in 2009 and caused 12,000 deaths in the United States. However, as vaccine rollouts occurred over time, these pandemics were minimized and kept under control. Still the distribution of health services in underdeveloped areas suffered from poor infrastructure, human skills, and equipment, leaving the very poor and vulnerable exposed to pandemics.

Across most societies, there was a gradual shift in public financing, lifelong learning, occupational choice, social mobility, demographics, economic inequality, values of nationalism, fear of falling behind, and mass illegal migration at border crossings. All contributed to a loss of self-confidence and rejection of public policy. This was the historical setting for the global status quo leading up to the pandemic beginning in 2020. Other long-standing concerns that faced global society from 2000 to 2019 included:

1. Bank security and fraud, internet hacking, and scamming through emails, mobile phones, telephones, and other modes such as *budol-budol* (a criminal *modus operandi* in the Philippines where a gang of accomplices entice or hypnotize an innocent person);
2. Offshore manufacturing of goods and services with extended supply chains, trade and finance, e-commerce, wealth and savings, circular and green economies, sunset and sunrise industries;
3. Food security, climate change, environmental protection, immigration, supply chains, international development, debt traps for poor countries, and access to services;
4. Sickness and health, family paid holiday leaves, social relations, the rise of influencers and followers; and
5. Personal freedom and self-confidence, fear of aging and mortality, poor individual human development, learning modalities, science and technology, and machine and human learning.

Each behavior, attitude, and activity was ripe for change when the pandemic swept across the globe starting 2020. Under these conditions, COVID-19 would cause many sectors, institutions, and personal points of view to change, thereby altering the *status quo* (see Figure 1.1).

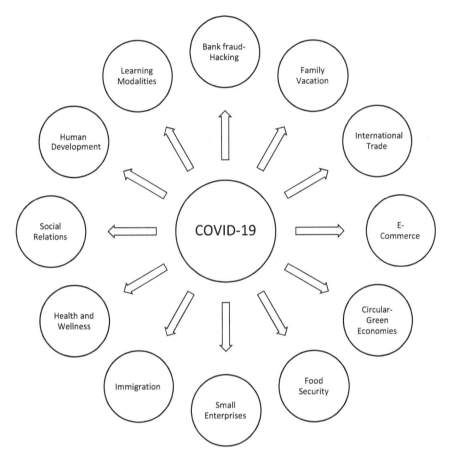

FIGURE 1.1 The many faces of change altered by COVID-19.

Source: © William Loxley and Suresh Nanwani

The first year of the pandemic (2020)

The pandemic came like a phantom in the night, causing sickness, death, and uncontrolled disruption throughout the world. The rise in infections, deaths, hospitalization, family grief, school closings, food shortages, and earnings decline were enormous. Transport shortages, work disruptions, curfew, stay at home, and quarantine restrictions with no visitors to homes became commonplace. Online internet versus offline face-to-face meetings; face masks and social distancing; restaurants, cinemas, entertainment, and sports venue closures; and routine life changed in 2020.

All these changes suddenly burst upon the global stage and caused fear and panic. Most obvious was the effect of lockdown on occupations. Whole sectors ceased to function: tourism and travel; restaurants, hotels, entertainment, sport, and fitness; shopping, manufacturing, transport, schools, and office work. These changes dramatically affected millions in employment, business, and leisure lifestyle in 2020.

In *Covid-19 in the Philippines: Personal stories* (Nanwani & Loxley, 2021), the 40 stories recounted by people from various walks of life demonstrated the spectrum of experiences and changes they experienced in the first year of the pandemic. Writers include a house helper, an engineer, a gardener, a seaman, a security guard, a student, a medical technologist, a professor, a nurse, a property management corporation director, an international development organization specialist, a school principal, and a priest. The book documents how people reacted to a social emergency that no one initially comprehended.

Epidemiology studies the incidence, distribution, and control of disease relating to public health. If there is no common response to command and control outbreaks, pandemics will flourish. WHO, a UN agency tasked with monitoring global public health, initially appeared slow to mobilize policymaking at the onset of COVID-19. WHO seemed not to take direct action, failing to alert and educate the world of the origins and dangers of the virus early on, in spite of having more information on the ground (WHO, 2021). When a virus jumps species, it is vital to trace how and when transmission occurred.

It was only in mid-January 2020 that WHO announced that human-to-human transmission was possible. By then, contagion was already rampant in places like Iran and Italy. The lack of international coordination fueled the early spread by January 2020, failing to slow down the spread and mobilize protocols to limit contagion. Thereafter, WHO appeared to handle the pandemic well.

Already rampant locally in Wuhan, China, by New Year's Day in January 2020, COVID-19 was not widely reported outside communities. As this did not give rise to concern, it caused a delayed international response to the rapid spread of the virus. In the meantime, global air travel spread the disease to Europe, throughout Southeast Asia, and the Americas, and less so in Africa, North Africa, and the Middle East. These latter regions were spared probably because they were less economically connected to world trade and travel. Australia, New Zealand, and later, China closed off their borders early with a zero-tolerance policy, which temporarily limited pandemic effects. The composition of the COVID-19 virus DNA was released by the Chinese government in early January 2020 when it was confirmed that the virus spreads to humans through the air, from nose to throat to lungs. From then on, WHO began to distribute information widely and quickly.

A word about vaccine types is in order (CDC, 2022). There are three main types of vaccine production: (i) use of a whole virus in an inactive state to make vaccines for personal injection; (ii) use of parts of the whole virus to trigger immune system response; and (iii) specific use of inactive virus mRNA strands similar to (ii) but using only minor genetic material found to be the most effective in triggering the immune system without causing infection.

Types (ii) and (iii) are called mRNA vaccines, both of which are relatively experimental and therefore had to undergo trial testing before release. Later, this approach will appear to be overall the most effective. It appears the second and third types of vaccines are more readily adapted vis-à-vis new variants of COVID-19, with relatively high levels of efficacy of around 90 percent. Of course, no vaccine guarantees complete freedom from catching COVID-19, but it is clear that those previously vaccinated experienced mostly mild symptoms and often did not require hospitalization, as is not the case for the unvaccinated.

A brief mention of the virus's symptoms, maturation, and trajectory is also in order. The *modus operandi* of the virus includes symptoms similar to the flu – fever, headache, sore throat, loss of taste and smell, cough, and difficulty in breathing. Worse cases involve the inability to breathe as the virus causes the lungs to fill with liquid. It also affects body organs such as the heart, brain, and kidneys. This causes sheer panic and results in body shutdown as breathing becomes difficult. The global response to the pandemic was like a replay of some Hollywood disaster movie scaring its audience, while other viewpoints claimed the pandemic was no worse than the typical flu.

In general, upon contracting the virus, one might carry it asymptomatically and feel fine, or fall sick. In either case one would be contagious. Isolating at home comes next, followed by admission to the hospital when breathing becomes nearly impossible. In the hospital, severe cases are intubated while in ICU. If one recovers from the hospital stay, one goes home to recuperate. If not, the person just dies.

Other issues related to COVID-19 include the severity of the virus and the need for social distancing, support for frontliners, and provision of financial assistance to the most needy. The amount and degree of exposure to the virus affect its virulence. Therefore, the use of face masks, handwashing, social distancing, quarantine, curfew, testing, and general avoidance of crowds are essential. Finally, pharmaceutical discovery of new vaccines and testing prior to release were amazing events often overlooked.

The general reordering of life toward staying quarantined at home to avoiding travel to work became commonplace and unsettling compared to normal routines. The idea of wearing face masks in public became novel and annoying but necessary for many; for others, it was an intrusion of privacy. The number of infections increased from personal exposure, putting pressure on hospitals and frontliners as ICUs filled, oxygen supplies dwindled, and deaths mounted. However, the face mask mandate slowed the progression of infections at a time when there was no vaccine protection. It made practical sense although some individuals were unable to adjust to wearing them, perhaps due to a fear of a mockingly unflattering self-image when looking in the mirror. In addition to face masks, face shields were required in the Philippines for the first two years.

All in all, economies shut down to limit the virus spread, leading to near total disruptions in daily routines. The advent of the pandemic introduced new terminology to the international community lexicon in 2020: face masks, face shields, quarantine, lockdown, COVID-19, frontliners, social distancing, contact tracing, swab tests, long-haul symptoms, herd immunity, booster shots, comorbidity, and vaccination ID.

Many people took time adjusting to social distancing protocols, including wearing masks and keeping a two-meter distance in crowded areas. The backlash to public health protocols was something unforeseen (Mercola & Cummins, 2021). Perhaps it was a fear of needle jabs frequently showed on TV that added to the unease. Later on, vocal resentment against COVID-19 protocols by a minority of anti-vaxxers surprised many, given the potential danger of the virus to the individual and community. It seemed the ideology of personal freedom versus following rules was at the heart of this opposition spread by media misinformation, coverage, and the like. A mixture of fake social media, fear, ignorance, uncertainty of pandemics, inequality, and poor political dialogue would require greater socialization emphasis to build trust in the common good.

One can add another reason why some communities engaged in vaccine hesitancy and suspicion of information on COVID-19. They did so due to not understanding science, which led to distrust of authority. Fear of foreign things entering the body made people think only about themselves and not the community. Rural areas usually give short shrift to COVID-19 than urban centers, given their lower population density and closer interaction with neighbors living in a relaxed lifestyle.

In 2020, schools were forced to close (some permanently) and hundreds of millions of students and teachers stayed home, not to mention parents who had to babysit children 24 hours a day.[3] As there were no face-to-face classroom sessions, children stayed with parents working from home. This required extra logistics for those who were working while teaching their children. Daily routine was unable to adjust quickly to the requirements of online learning. As a result, academic skills declined as computer gaming time increased (Netswera et al., 2022).

Stay-at-home options caused work from home to rebalance employer-employee relationships as productivity had to be recalculated and verified versus wage premiums paid to some. Office rentals declined along with ancillary support systems, including restaurants, transport, and small-scale services.

Finally, home life with family members was greatly altered.[4] During quarantine in close quarters, interpersonal relations frayed: Children longed to play outside with friends; parents squabbled over chores and finances. Children, parents, and seniors were all confined to close quarters for long periods.

This took its toll on their sense of purpose, sometimes causing depression and aggression. Imagine seniors living out their final years, wondering if the end will come much sooner due to catching COVID-19.

These examples present a mental picture of lifestyle changes that occurred over several years under COVID-19. All the while, policymakers were caught up in the global unhappiness syndrome, unprepared to mobilize public opinion and educate populations while having to manage the pandemic (hospitals, testing, quarantine, welfare transfers, procurement of medical supplies, and allowances for frontliners).

Figure 1.2 tracks the two biggest shifts in time spent on activities between 2019 and 2020 that shaped values and behaviors on social, physical, mental, psychological, and emotional well-being. The time-tracking in Figure 1.2 is based on Cassie Holmes's book, *Happier Hour*, which looked at the time spent meaningfully in work and play during the pandemic. Although the percentages are rough estimates, the figure shows time in office and school declined while family time increased. Likewise, offline face-to-face activities such as sports, concerts, and social clubs declined dramatically while online virtual connections to gaming, sing-along, Netflix (Baidu in China), and social media expanded rapidly. These two major shifts in daily time spent on activities, mostly in homes, need to be reviewed carefully to see if trends will remain or revert to the old normal. At the same time, schools remained on the periphery of social action due to the pandemic.

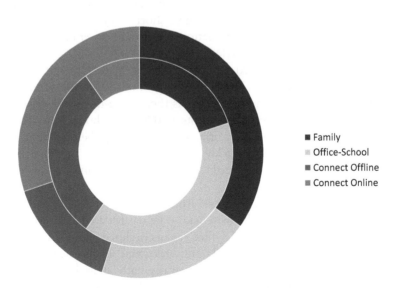

FIGURE 1.2 Time spent on activities (outer circle 2020, inner circle 2019).

Source: © William Loxley and Suresh Nanwani

In advanced nations, deaths among seniors, the fragile, the poor, and the homeless were severe, as even relatively good hospitals could not cope with the high number of the infected. In poor nations with inadequate health care systems, the problems were multiplied severalfold. This was the picture of 2020 in an era of fear and panic that swept across the globe.

The second year of the pandemic (2021)

In retrospect, it is easy to see what was happening by 2021. COVID-19 was again on the rise in the cold climates of Europe and North America during the first quarter of the year. By April, a new Delta variant wave of COVID-19 swept through India and South Asia and elsewhere, causing wide infection and deaths even among the young. At the same time, vaccines were being provided for free, mainly to citizens in advanced countries. The standard protocol offered two doses administered over one to several months, depending on the vaccine. It turned out that the vaccines worked well for both the original Alpha and new Delta variants. However, as most of the world is poor, both urban and rural dwellers still lacked immunization by the end of 2021.

By the last quarter of 2021, yet another wave of COVID-19 named Omicron spread first from South Africa into Europe and North America. By the end of the year, it spread well across Asia and beyond. Although the Omicron variant spread rapidly, it was not as virulent as the Delta variant. It seemed most mRNA vaccines worked well enough to offer reasonable protection against it. This was especially true for the vaccinated, whose global numbers swelled on average to one in three in each population.

By late 2021, Omicron had already spread across the world. However, it began to decline by late March 2022. By then, most of Southeast Asia, Europe, and the Americas had infection rates under control, especially among the vaccinated. Hospitals were not overwhelmed, and economies no longer needed to be totally shut down.

All in all, by the end of 2021, after two years of long-running lockdowns and quarantines, the pandemic had diminished in most places, flaring up only among the unvaccinated and places possibly not previously exposed to COVID-19. For instance, many Pacific nations remained unaffected by COVID-19 until August 2020, given their geographic isolation. The next year, however, these nations finally succumbed to the pandemic, such as Tuvalu in October 2021, when it recorded its first case of COVID-19.

Deaths were not quite as severe as in 2020, and by late 2021, vaccination programs had been administered to well over 60 percent in populations of advanced nations, thereby slowing infection rates significantly. Still, by the end of the second year, the world had lost approximately 6 million residents, with over 500 million infected and recovered, and perhaps as many as 10 million or more living with permanent damage to the lungs, heart, and kidneys.

In 2021, major disruptions occurred in the world of work, with shutdowns costing trillions of dollars in lost revenue. School closures caused learning deficits in 2020 and 2021, experienced especially among the disadvantaged who could not afford computer and internet service to follow online study. Other institutions severely impacted by COVID-19 included:

- hospitals with frontliners who either perished or suffered severe mental and physical fatigue from the demands of emergency room services during the height of the pandemic;
- transportation shutdowns that prevented people from getting essential services, which caused social service disruption to government assistance programs;
- medical science research that barely managed to successfully develop and test a new mRNA virology technology to create relatively safe and effective vaccines;
- public agencies, NGOs, and small businesses that struggled to provide medical and clinical staff with the necessary safety equipment; testing, monitoring, and tracking of infected individuals to prevent the spread of new and more deadly variants caused by new subvariants;
- the lack of cultural venues left only a few ways to creatively debate how social conventions and pandemics converge to aid or disrupt social progress; and
- the loss of income-generating foreign trade and tourism.

Other changes included seniors losing self-confidence and self-esteem from a fear of dying. Uncertainty in climbing the career ladder affected youth, new entrants, and those already employed, given the shutdown. Healthcare options were impaired for those with severe COVID-19 comorbidities and others who died from other causes because hospitals could not accommodate non-COVID-19 patients. Children were unable to play with classmates and had to learn to study online. In general, a sense of indifference and lethargy pervaded individual thinking after enduring two years of on-and-off lockdowns, quarantines, and curfews.

In three words, 2021 was like Queen Elizabeth II's *Annus horribilis* speech in 1992: "another missed year" (just like 2020). Yet lessons were learned on how to adapt, especially as a result of new vaccination programs that limited the worst outcomes experienced in 2020.

The third year of the pandemic (2022)

The year 2022 started with severe Omicron infections in the first quarter, which slowly declined in the northern hemisphere and thereafter in the southern hemisphere.

With additional funding, WHO adequately monitored COVID-19 and even projected the pandemic to fade out in the last quarter of 2022, when it was thought to revert to an epidemic status. Latin America, the Middle East, North America, Europe, South Asia, and Southeast Asia appear to be adjusting to the new COVID-19.

Two areas remain in question and could incubate new and virulent strains if not monitored carefully and reported to WHO. The African continent has over a billion people, but less than 15 percent are vaccinated. Fortunately, COVID-19 infections appear limited in Africa, perhaps because of its young population, with half of everyone under the age of 20. Perhaps fewer infections are reported because the health system simply does not engage sufficiently with the poor to report statistics. In addition, large swaths of the population live in rural areas, which are less densely populated and less likely to spread COVID-19. Should a new variant arise in Africa, it would likely spread quickly around the world, as did the Omicron variant that started in South Africa.

The second area of concern is China. With over a billion people, most of the population is vaccinated against the earlier COVID-19 strains. But not many are vaccinated against Omicron, given the limited use of mRNA-type vaccines in China, which work somewhat better than traditional vaccines to control Omicron. As a consequence, this large population could generate new forms of COVID-19, which could spread across the country and the world, thereby keeping the pandemic from disappearing. Given its population of over a billion, India probably should also closely monitor its rural population for signs of new variants springing up and spreading to urban areas.

During the first half of 2022, COVID-19 continued to remain a pandemic. Thereafter, no new major variants emerged that could not be controlled by existing vaccines. It would appear that the COVID-19 pandemic may move to endemic status and no longer threaten day-to-day global social, economic, political, and cultural activities. The reopening of air travel, religious and festive holidays, elections, vacations, and business trips in late 2022 have not increased infection rates dramatically. However, death rates should remain low given high vaccination rates. The promise of oral vaccines may also make it easier to stop the spread of infections.

To keep current population vaccine coverage at the minimum of 70 percent and above, there needs to be continuous vaccination programs, especially for the vulnerable subpopulations, including seniors and children. Because new variants of COVID-19 might emerge anytime, populations may need to receive regular or annual booster shots to keep the virus from turning into a pandemic again. Nevertheless, scientists now know that although the virus is more dangerous for specific groups, they can be safeguarded effectively with vaccines and social distancing protocols.

Ending the pandemic in 2023 depends on three factors:

1. No new deadly and virulent COVID-19 strains arise;
2. Old and new vaccines remain potent over reasonable time, and frontliners, seniors, those with comorbidity, and other at-risk workers and subpopulations get booster shots; and
3. Governments and citizens maintain vigilance in supporting preventative public health measures to keep small outbreaks at bay.

Three years of fighting the pandemic have shown that it is difficult and costly to track and trace (in order to stay ahead of an out-of-control pandemic) and command and control (to manage a quarantine). There are simply too many ways for infections to cross barriers in modern society. Thus, there needs to be some balance between lockdowns and business as usual. Take the United States as an example of the trade-off between public health and the economy. In the beginning, they tried a *laissez-faire* approach, which did not work. The second phase was severe lockdowns and monitoring (even in a society of relatively strong public health), which barely kept pace until vaccines became readily available. The third phase continued social distancing along with strong vaccination and monitoring, which allowed the country to open up the economy. This became the model for many countries trying to balance pandemic safeguards with resuming daily life.

The pandemic adjustments in 2023 and beyond

In general, the pandemic lockdown has affected nearly eight billion lives globally in many ways. Permanent changes to institutional reorganization affected many, as businesses ramped up e-commerce and home delivery, and social media apps moved into gaming activities and online social discourse. Permanent changes also occurred in financial investing and payments, daily routines in school learning, work arrangements, recreational travel, value priorities, and health and wellness priorities linked to mortality (Bryant et al., 2022). Interpersonal relations, the rise of virtual reality, knowledge sharing through the internet and metaverse access, and early retirement options all figure into day-to-day thinking about the future. In all these cases, adverse life changes occurred when face-to-face interaction ceased to be an everyday option during the pandemic.

The start of the fourth year of the pandemic remains a question mark as to how long COVID-19 will continue or peter out. The latter half of 2022 suggested better control through more universally administered new forms of vaccines and medical breakthroughs, along with better ways for the economy to function in the presence of the waning pandemic. In May 2023, WHO declared that COVID-19 no longer represented a global health emergency.

Dr. Mike Ryan from WHO's health emergencies program cautioned that while the emergency may have ended, the threat was still there.

A nation's capacity to overcome adversity

Given the severity and spread of COVID-19 around the world, there is a need to employ a practical framework to capture the many issues facing society when emergencies suddenly emerge. Common sense tells us that four major sectors (social, political, economic, and cultural) in society provide overall support for national progress. These sectors carry out functions that enable people, power, money, and wisdom to self-organize and flow back and forth among people when conditions are favorable. These sectors nurture population welfare, power sharing, investment funding, and sharing ideas. Each sector gets organized and modernizes around upward social mobility, trust in government, economic productivity, and cultural creativity that builds technological innovation.

The book employs a structural approach to societal organization grounded in systems analysis useful in real-world social system planning (Easton, 1965; Babones & Chase-Dunn, 2012; Ryan, 2018; Rutherford, 2019; Mattani, 2020). Systems analysis is similar in approach to structural-functionalism in sociology (Turner, 2000). As such, the underlying concept of this study is more a method of research rather than an explanatory model. The analysis does not include a theoretical orientation with hypotheses to explain COVID-19 events. Rather the investigation suggests future analysis relating COVID-19 to society because there is no comprehensive model at present that explains social dynamics of the virus on the world. Analysis can focus on the structure of societies and the actions taken by institutions to function in a pandemic. The approach helps describe and think about how society adjusts to internal and external threats to it.

When apocalyptic events do occur, these sectors respond to minimize disasters by keeping nations functioning. There is an enhanced probability of success in fighting pandemics when opportunities for social mobility increase commitment to learning, career, public trust, and technology through cultural values and behaviors. At the same time, the political economy reinforces the rule of law as it applies to fairness and the common good. Together these processes represent a unique prism to view society as they alter the *status quo*. The more these processes work together, the greater and faster dynamism counters old ways. With this framework in mind, key activities facing each sector can be identified, listed, and tracked to see how issues get resolved from solutions meeting current needs. Table 1.1 outlines these relationships that provide a context for exploring the COVID-19 impact on humanity.

Systems analysis is used to conceptualize socio-phenomena in many interdisciplinary arenas. Topics bring together in systematic and generalized form a conceptual scheme for the analysis of the environment, functional structure, and process. Likewise, structural analysis proposes models that apply boundary conditions reflecting the scope of function of individual parts to analyze and apply findings where feasible. Systems analysis studies core parts to identify objectives and shows how parts of the system work together. Core macro-sectors include social, political, economic, and cultural aspects of a nation. Systems analysis with roots in sociology examines how complex systems adapt and evolve in political analysis, organizational psychology in the workplace, cultural anthropology, and comparative international development.

This societal structure exists everywhere in the modern world to varying degrees, where each part contributes to the whole whether in crisis mode or stable periods. Society constantly reorganizes itself to accommodate stability and change through sector functions needed to keep it in equilibrium. For example: (i) education generates opportunities; (ii) public policy strengthens governance; (iii) occupational structures accommodate productive talent; and (iv) culture spreads information and values.

Systems theory is useful in (i) integrating the social sciences, similar to what Newton did for the physical sciences or what Max Weber's components of power did for the social order; (ii) examining the roles of sector functions in society and integrating them into a whole; and (iii) promoting socialization of education's role in reproducing knowledge, culture-defining society, governmental support for the rule of law, and business organization design. Common sense, sociological theory, and in this case, what the essays and news media found occurring in each sector of society over the years of the pandemic – all suggest key themes to emerge covering school learning, public health policy, organizational design of the workplace, and technical innovations and behavior.

The framework in Table 1.1 can be used to describe how societal norms help identify which movements in each sector succeed and which do not. Compared to 2022, how did Artificial Intelligence, data analytics, and the internet influence the youth prior to 2019, when society was less dependent on digital technology? Likewise, demographics, resource equity, technology, human capital, information on social media, fake news, minimal face-to-face exposure, and social etiquette conventions were all impacted by the pandemic. Each activity and event fits within the four-sector framework.

Institutions usually support society by widening opportunities for advancement, advocating fair play, distributing public goods, and debating standards of behavior (Halsey et al., 1997). The government provides political context, while businesses provide economic resources that offer opportunities to get ahead financially and save for the future. Families aid procreation and set out common values, while cultural institutions in the arts and sciences provide human creativity that gives direction to society. Digging deeper into this framework, we find the following four propositions.

TABLE 1.1 Structural elements and key sectors that promote change

Structural Elements	Key sectors			
	Society	Polity	Economy	Culture
Resource	People	Power	Wealth	Wisdom
Organizer	Social stratification	Laws and institutions	Marketplace	Philosophical ideas
Influencer	Individual networks	Stakeholder participation	Entrepreneurial investment	High-tech approaches
Transforming Mechanism	Social mobility	Trust	Productive work	Creative thinking

Proposition 1: The basis of society is its people. The population's wellbeing, along with the geographic distribution of the individual's groups, is stratified by levels of attainment. Social relations allow nations to organize social networks (Fagerlind & Saha, 1981). The more fluid a society, the more choices and the greater social mobility for people. For example, when the COVID-19 pandemic started, schools closed suddenly and young people found themselves isolated from the traditional paths of going from school to work to family formation, and later to lifelong learning. Skill training stopped, along with job hiring and postponed weddings, leaving the younger generation with a sense of falling behind. Because social aspects of human organization reside within families and small groups, these structures serve as the means for people to advance from one stage to another in the human life cycle. The social structure helps ensure a smooth transition from school to work to family in each generation. One may ask whether COVID-19 has altered values and behaviors, especially among the next generation in pursuing alternative routes to success (Ryan & Nanda, 2022).

Proposition 2: The basis of the polity is power-sharing consensus. Its structure is based on control through institutions and rules that confer legitimacy. The flow of political power lies in individual participation in civic responsibility where good governance engenders trust and educated citizens can weigh alternatives (Haugaard et al., 2012). When the COVID-19 pandemic started, little was known about how to manage it. Governments had to figure out protocols such as testing, isolating, tracking, social distancing, and social protection. They then had to explain the science to the public and encourage public participation to support programs (Dipankar, 2021).

Governments had to co-op divergent groups offering alternative information on how to confront and control the spread of the virus, especially across regions and among those poorly protected against public health disasters. The political sector revolves around trust in the common good to which the vast majority aspires based on common values learned in schools.

Proposition 3: The basis of the economy is wealth. Wealth is structured through the market place by buyers and sellers who meet public demand (ILO, n.d.). Change in the economy results from productivity gained through investment over savings, which leads to higher GDP. The economy provides employment to carry out work in large and small businesses. All effort demands large amounts of time and money spent on national development designed to keep the wheels of prosperity turning in offices and workshops. For example, under the pandemic, remote work or work from home (WFH) became mandatory for most, thereby disrupting the entire office and business culture.

Under extreme lockdown and quarantine, offices, factories, businesses, and the marketplace shut down and people worked from home. Without vaccines, society was left with no choice but to shut down the economy to prevent the public health system from collapsing and elderly population from being decimated. This event forced the sector to reconfigure the work and career experience in major ways that may not be easy to undo. Occupational change caused long-lasting problems with values toward the workplace, career advancement, and balancing work-home life through socialization skills taught in school. The economic sector, from big companies and small businesses, also had to find ways to keep everything afloat in a time of low demand.

Proposition 4: The basis of culture is collective wisdom (Bruner, 1997; Resette-Crake & Buckwalter, 2023). Culture alters the collective conscience of a nation that gives purpose and choice to adapt and express values and behaviors. Education is inseparable from culture, as technology spreads ideas. Culture encourages diverse thinking in language, cuisine, fashion, and epistemology, leading to creative lifestyles (Bouronikos, 2022). Creativity is the force that conceives ideas and methods, which give new meaning to old ways. Culture expands through religious organizations, schools, concert halls, lectures, and exhibit venues, all of which contribute to dialogues among groups to reach a common understanding. Culture offers opportunities through contact with interesting people, useful practices, and new ideas. Unfortunately, under quarantine, people became socially isolated and lost the ability to communicate intelligently. Even WHO designated computer gaming as a disease as young people began spending more time gaming and gambling at the expense of human interaction.

During the pandemic, the internet became the main driver of daily cultural activities and social contact, especially among the youth, business people, seniors, mothers, and children under quarantine and lockdown (Gammel & Wang, 2022). To the extent that overreliance on the internet leads to dependence on technology, daily life will reinvent values and behaviors. During the pandemic, sector activity ceased to function as everyone went on social media to blog, tweet, and zoom to meet and vent frustrations.

This isolation of face-to-face content for pre-teens, youth, and the elderly became dramatic over time as community interaction ceased. Family abuse increased during COVID-19 lockdown. The switch from real to virtual reality may impact the way entire generations receive and evaluate information, engage in recreation, financial investment, shopping, and more. Changes in human values and behaviors affect cultural norms and alter the way people think.

Daily activity that promotes well-being for most citizens without harming others is the goal of well-balanced societies. Most open societies continuously discuss values and attitudes to support the general well-being of their people, including diversity, inclusivity, and other traits needed to smoothen the flow of progress. Even protest movements are often a good sign that society is well balanced, as they encourage debates that address minorities, workers, and women's issues. As the fundamental needs of society to cooperate over individual choice allow common interest to take hold, the pandemic clearly opens debate about cooperation and debate (Novak & Highfield, 2012).

Drawing on the aforementioned framework, this book uses sector perspectives to organize social action and reaction that gives meaning to progress. Certain events in specific areas may move more slowly and unevenly than others while other areas may move more rapidly. Yet, in the long run, human development seems to advance. In this context, COVID-19 suddenly appeared and forced changes in the way society operates and perceives progress. Some individuals improved their lives, while others stagnated and sank into depression. At this juncture, we invite readers to revisit Table 1.1 to appreciate and better understand how COVID-19 affected national development around the world.

Exploring COVID-19 influences on society

Societies and civilizations often find ways to package daily life efficiently into connected activities that sustain and improve the well-being of citizens. This collective approach combining social, political, economic, and cultural dimensions is one way of framing such a mammoth task of coordination, especially in times of crisis (Hanafi, 2022; Hattke & Martin, 2020). Dynamic cultures struggling in the midst of out-of-ordinary disasters must put the COVID-19 pandemic experience into context so people can understand its impact on life's journey. This challenge requires information dissemination, education, and training across all sectors of society.

Readers are encouraged to ponder on how their personal experiences under COVID-19 influenced many following interconnected activities that go into forming a vibrant and resilient society. Everyone has a life-altering experience from which they can apply their own expertise and experience to explain how COVID-19 changed common lifetime features. While Table 1.2

TABLE 1.2 Four key issues in national sectors arising from the pandemic

Social issues	Political issues	Economic issues	Cultural issues
Social mobility	Vaccine distribution	Job-family balance	Values behaviors
Education attainment	Public policy	Financial literacy	Self-confidence
Skills training	Public participation	Remote work	Personal freedoms
Cognitive skills	Law and order	Small businesses	Self-actualization
Emotional IQ	International trade	Office-factory life	Human development
Mental wellness	Information flow	Wages	Lifelong learning
Personal aspirations	Geopolitical rivalry	Access to justice and dispute resolution	Compassion
Social networking	WHO leadership	Climate change and environment	Cooperation
	Virtual courts		Human connectivity
	Online degrees		Cultural technology

itemizes these life episodes by core activity, readers are invited to add their own to the list.

The social sector and educational opportunity

The main impact of COVID-19 on the social sector appears to be personal isolation and a lack of all forms of day-to-day mobility. This resulted in people having the impression of downward social decline. The lack of opportunities limits upward movement and increases downward direction especially for the poor. All forms of face-to-face meetings in religious organizations, schools, offices, concerts, and meetings were curtailed, leaving mostly contact via the internet, which limited personal friendships. Keeping up with the latest trends and fashions and new ideas forced people to go online to watch influencers and shop on e-commerce companies such as Amazon, Lazada, and Shopee.

Isolation resulting from a lack of mobility limited the perception of upward movement in society. Learning deficits resulting from a lack of face-to-face contact and overreliance on the internet way of learning contributed to apathy and a loss of control over the future. Without personal contact, people were locked into mostly passive paying, clicking, and watching content for adults. For the youth, it was "eat, sleep, and computer gaming." Fewer daily real-life experiences led to fewer opportunities to develop socially. In the case of education, school closures and early teacher retirements led to a loss of learning. Since schools are the fulcrum of a well-functioning society, discounting them puts the whole nation at risk.

The political sector and managing public policy

A major dilemma facing the political sector during COVID-19 was how to organize and pay the mobilization costs of the pandemic (Mossleh, 2022).

This is especially true for the public health and social welfare of those who are most in need. These expenditures were made to reduce hardships for the poor during the pandemic. Governments also struggled to disseminate information and organize various aspects of COVID-19 testing, contact tracing, vaccination, and provision of money transfer allowances for food and rent to the needy (Duca & Meny-Gilbert, 2023).[5] The public sector had to coordinate the purchase of medical equipment for hospitals and frontliners, and later, vaccination supplies of vaccines and needles. Staff remuneration also totaled in billions of dollars worldwide.

According to WHO, funds for vaccines and welfare transfers to the poorest households due to work redundancy caused by COVID-19 amounted to USD 17 trillion. This included lost productivity due to the economy shutting down (Reuters, 2022). Governments had to be careful to inform the public on policy decisions if they wanted to maintain trust in supporting the trade-off between economic shutdown and health safety. Political activities tied to cash disbursements to the indigent are always sensitive. Nonetheless, balance had to be maintained even when it was not clear how successful protocols would be in reducing infections or how painful economic lockdowns would affect the poor. In this case, vaccines were game-changers in reopening economies around the world. We predict this issue will be up for continuing debate when deciding the best approach to address future disasters. Most governments might be reluctant to implement nationwide lockdowns for entire economies in the future.

The economic sector and learned skills and attitudes toward work

During the pandemic, the economy was tasked with continuously providing society with goods and services. There was a shift from services to goods as people bought items for the home to enjoy during quarantine and curfew. This meant less money was spent on services and more on the consumption of home appliances and in-home dining. Supply chain disruptions resulted from the move to e-shopping and dining-in as home delivery dramatically increased. The old-style department store was forced to reduce staff and cut off suppliers as shoppers declined. Many small businesses, including restaurants and retail stores, struggled as demand collapsed while owners tried to keep staff working from home.

Changes in the workplace resulting from office lockdowns forced employees to work from home (WFH). This ultimately strengthened worker bargaining positions with employers once offices and stores reopened.[6] No wonder many employees did not want to return to 9–5 workdays in the office or the shop floor when they had more control over how they balanced their work and family lives. In addition, WFH lessened the amount of time employees spent at work with colleagues and increased their time on childcare, moonlighting, and pursuing personal tasks on company time.

The cultural sector and innovative well-being

The cultural sector relies on human ingenuity in juxtaposing day-to-day routines with new perspectives on how and why humans behave the way they do to survive pandemics. Media consensus from data and surveys suggests that under COVID-19, society became less dynamic, less energetic, and less resilient, with more depression and violence in families (OECD, 2020). Closed venues, less innovation, and little personal contact led to the loss of cultural continuity across generations, as sociopsychological life forces fragmented.

With little stimulating social contact during the pandemic, artists and intellectuals were left to their own devices on finding stimulants to reconstruct reality without feedback over coffee at a café with fellow artists.[7] The pandemic caused culture to question ideas like fairness on a host of topics, including women, minorities, workers, and the poor, in terms of pay and access to health services and childcare. All the while intellectuals lacked direct access to real-life difficulties and experiences. Politicians and captains of industry also lost control of everyday challenges facing constituents or the business cycle.

Society also had to rethink its relationship to planet-friendly sustainability. For example, the transformation to near-zero emissions requires a green circular economy that is likely to eliminate and create tens of millions of jobs over the coming decades. Finally, COVID-19 caused society to depend more on the internet for entertainment, learning, work, and social communication than in the past. The new information dissemination model made it difficult for the public to separate fact from fiction in knowledge sharing, given the anonymity of those posting content on blogs.

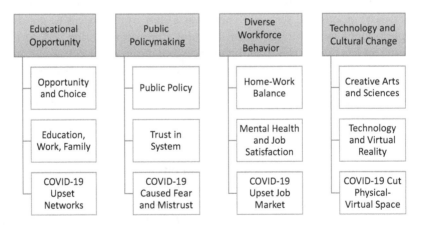

FIGURE 1.3 Covid-related central themes emerging from social, political, economic, and cultural sectors.

Source: © William Loxley and Suresh Nanwani

Based on what has been already learned and debated about pandemic effects over the past three years, key themes have emerged relevant to the social, political, economic, and cultural domains (see Figure 1.3). These themes reflect approaches to life: educational opportunity for social mobility; informed public awareness; career guidance; and technology for self-improvement.

Themes that might alter post-pandemic relationships

Some human activities have been greatly affected by COVID-19 more than others. These areas are likely to recover from the economic shutdown after two to three years, perhaps in a different format. Keeping in mind sociology's role in tracking global development, the following key areas are identified:

- **Demographic shifts:** The pandemic taught the world that senior citizens are highly vulnerable to health problems and more susceptible to viruses given their relatively weaker immune systems. As the world continues to age in the 21st century, this lesson has wide implications for planning healthcare and servicing needs of the elderly, which would not otherwise have registered in the public psyche. In the case of non-seniors, taking away the mobility that leisure travel affords left a big hole in many lifestyles, which led people to fear a loss of control and end up being depressed or suicidal. Midlife career change and early retirement also affected the workforce. Likewise, those with comorbidities like heart, lung, kidney, and liver problems all face additional issues, limiting their life expectancy every day.
- **Science research and innovation:** Accelerating science research and innovation, especially in medicine, has benefited from the pandemic. An example of this is the rapid creation, production, testing, and delivery of new vaccines within a short time span. Those who fail to appreciate the benefits of science during the pandemic are doomed to not see the value of science elsewhere. Research on vaccines, health protocols, and telemedicine are areas where science supports future health systems effectively.
- **Social mobility:** This results from wider opportunities for careers due to lower unemployment rates. Early retirements of older workers, new protocols for work from home, gig-style work outside offices, and the youth's alternate valuations of traditional career and family formation steps – all of these indicate rising social mobility options for many, especially young people.
- **Life-choice experiences:** This requires fairness based on personal fulfillment in the way parents and children see the future. The way individuals and society come to value everyone – from Wall Street traders to hotel room cleaners – is a challenge, as many occupations remain segregated from each other. This perception of differences in occupational value held by people forces society to assign importance to some more than others, causing depreciation in valuing hard work based on job description.

- **Life-long learning:** The process facilitates many avenues for gathering knowledge and information. When restricted by a lack of internet service or ICT software and apps, people can feel alone and isolated. Staying home during the pandemic has left people highly dependent on television, YouTube, Facebook, TikTok, gaming devices, and entertainment apps. When service disruption occurred in the digital sector, loss of connection to the outside world became unbearable to many. This confirmed people's overdependence on the internet in daily life.
- **E-internet information technology (IT):** IT and the metaverse include all aspects of email, social media, gaming, zooming, blogging, video production, and book publishing.[8] They represent a slow-moving, continuous switch from reality to virtual digitalization sped up by the COVID-19 lockdown that kept people dependent on digital platforms. Technological connectivity will have a large impact on altering the way people think and behave.
- **Climate change and environmental protection:** As shown in satellite pictures, the two-year shutdown has dramatically limited carbon dioxide emissions and allowed the world to contemplate on how modern living has adversely affects air pollution, weather swings, and climate and resource degradation in the Amazon basin, Himalayas, Levant, and elsewhere.
- **International trade:** COVID-19 has altered development aid, protectionism, supply chains, foreign exchange rates, factory closures, migrant labor and capital flows, technology waves, and trade in services. As world trade reconnects, nations will become more wary of depending on long-linked supply chains.

Surveying COVID-19 consequences on global development

Having considered the wide impact of COVID-19 on the way social mobility encouraged societal development, we next drill down to specify how outcomes of the pandemic shaped the revised normal (Rashawn & Rojas; 2020; Sachs et al., n.d.). Societies do not have to return to the old normal.[9] Chapter 2 provides a social survey based on a case study approach, plus the aforementioned themes and topics, through essays explaining how individuals around the world have adapted to the COVID-19 experience from 2020 to 2022. These essays tell about the changes brought by COVID-19 during the pandemic and the changes thereafter.

Notes

1 Most mainstream media reported similar assessments of the ongoing pandemic, and various research organizations and search engines echoed these results.
2 See Chapter 18, "The goldfish and the net(work)" by Nicole Mazurek.
3 See Chapter 7, "Adapting to virtual education during the COVID-19 pandemic in Peru" by Victor Saco.

4 McKinsey & Company reveals that many workers no longer want a traditional position with traditional pay and perks. The tracking, including a survey of more than 13,000 respondents in six countries, shows that many people are reevaluating what they want from a job and from life. They want more flexibility, mental health support, and meaningful work.
5 See Chapter 16, "Change and continuity: COVID-19 and the Philippine legal system" by Antonio G. M. La Viña.
6 See Chapter 23, "COVID-19 and moving to the new normal" by Victoria Márquez-Mees.
7 See Chapter 30, "I'm gonna let it shine: local musicians in the Virginia countryside" by Lori Udall.
8 See Chapter 27, "Will technology replace or recreate humans" by Arthur Luna.
9 Not returning to the old normal but moving to the "next" or new normal suggests a sense of change and progress.

Reference list

Babones, S., & Chase-Dunn, C. (2012). *Handbook of world systems analysis*. Routledge.
Bouronikos, V. (2022, April 22). *Culture and creativity are key elements for sustainable development*. Institute for Entrepreneurship Development. https://ied.eu/blog/culture-and-creativity-are-key-elements-for-sustainable-development/
Bruner, J. (1997). *The culture of education*. Harvard University Press.
Bryant, J., Child, F., Dorn, E., Hall, S., Schmautzer, D., Kola-Oyeneyin, T., Lim, C., Panier, F., Sarakatsannis, J., & Woord, B. (2022, April 4). *How COVID-19 caused a global learning crisis*. McKinsey & Company. www.mckinsey.com/industries/education/our-insights/how-covid-19-caused-a-global-learning-crisis
Center for Disease Control and Prevention. (2022, September 16). *Understanding how COVID-19 vaccines work*. www.cdc.gov/coronavirus/2019-ncov/vaccines/different-vaccines/how-they-work.html
Dipankar, S. (2021). *Pandemic, governance and communication: The curious case of COVID-19*. Routledge.
Duca, F., & Meny-Gilbert, S. (2023). *State – society relations around the world through the lens of the COVID-19 pandemic: Rapid test*. Routledge.
Easton, D. (1965). *A framework for political analysis*. Prentice-Hall.
Fagerlind, I., & Saha, L. (1981). *Education and national development*. Elsevier. www.elsevier.com/books/education-and-national-development/faegerlind/978-0-08-030202-7
Gammel, I., & Wang, J. (Eds.). (2022). *Creative resilience and COVID-19: Figuring the everyday in a pandemic*. Routledge.
Halsey, A., Lauder, H., Brown, P., & Wells, A. (1997). *Education, culture, economy and society*. Oxford University Press.
Hanafi, S. (2022). *A sociology for a post-COVID-19 society*. Bristol University Press.
Hattke, F., & Martin, H. (2020). Collective action during the Covid-19 pandemic. *Administrative Theory and Praxis*, 42(4), 614–632. https://doi.org/10.1080/10841806.2020.1805273
Haugaard, M., & Ryan, K. (Eds.). (2012). *Political power: The development of the field*. Verlag Barbara Budrich. www.amazon.com/Political-Power-Development-Science-Discipline/dp/3866491050
Holmes, C. (2022). *Happier hours*. Simon & Schuster.

International Labour Organization. (n.d.). *Economic and social development*. Retrieved December 24, 2022, from www.ilo.org/global/topics/economic-and-social-development/lang-en/index.htm

Kharas, H., McArthur, W., & Ohno, T. (2020). *Breakthrough: The promise of frontier technology for sustainable development*. Brookings Institution Press.

Knight, K., & Bleckner, J. (2023). No magic bullets: Lessons from the COVID-19 pandemic for the future of health and human rights. In J. M. Ryan (Ed.), *COVID-19: Individual rights and community responsibilities*. Routledge.

Mattani, M. A. (2020). *Cultural systems analysis*. Springer Publications.

Mercola, J., & Cummins, R. (2021). *The truth about COVID-19: Exposing the great reset, lockdowns, vaccine passports, and the new normal*. Amazon Kindle Books.

Mossleh, P. (Ed.). (2022). *Corona phenomenon: Philosophical and political questions*. Brill.

Nanwani, S., & Loxley, W. (Eds.). (2021). *Covid-19 in the Philippines: Personal stories*. Amazon Kindle Books.

Netswera, F., Woldegiyorgis, A. A., & Karabchuk, T. (Eds.). (2022). *Higher education and the COVID-19 pandemic: Cross-national perspectives on the challenges and management of higher education in crisis times*. Brill.

Novak, M., & Highfield, R. (2012). *Super-cooperators: Altruism, evolution and why we need each other to succeed*. Free Press.

OECD. (2020, September 7). *Culture shock: COVID-19 and the cultural and creative sectors*. www.oecd.org/coronavirus/policy-responses/culture-shock-covid-19-and-the-cultural-and-creative-sectors-08da9e0e/

Rashawn, R., & Rojas, F. (2020, April 27). Covid-19 and the future of society. *Contexts, Sociology for the Public*. https://contexts.org/blog/covid-19-and-the-future-of-society/

Resette-Crake, F., & Buckwalter, E. (Eds.). (2023). *COVID-19, communication and culture: Beyond the global workplace*. Routledge.

Reuters. (2022, January 21). *IMF sees cost of COVID pandemic rising beyond $12.5 trillion estimate*. www.reuters.com/business/imf-sees-cost-covid-pandemic-rising-beyond-125-trillion-estimate-2022-01-20/

Rutherford, A. (2019). *The systems thinker*. Verband Deutscher Zeitschriftenerleger (VDZ).

Ryan, J. M. (Ed.). (2018). *Core concepts in sociology*. Wiley-Blackwell.

Ryan, J. M. (Ed.). (2023). *COVID-19: Surviving a pandemic*. Routledge.

Ryan, J. M., & Nanda, S. (2022). *COVID-19: Social inequalities and human possibilities*. Routledge.

Sachs, J. D., Karim, S. S. A., Aknin, L., Allen, J., Brosbøl, K., Colombo, F., Barron, G. C., Espinosa, M. F., Gaspar, V., Gaviria, A., Haines, A., Hotex, P. J., Koundouri, P., Bascuñán, F. L., Lee, J. K., Pate, M. A., Ramos, G., Reddy, K. S., Serageldin, I. . . . Michie, S. (Eds.). (n.d.). The Lancet commission on lessons for the future from the COVID-19 pandemic. *The Lancet, 400*(10359), 1224–1280. https://doi.org/10.1016/S0140-6736(22)01585-9

Turner, B. S. (2000). *The Talcott Parsons reader*. Basil Blackwell Inc.

Turner, B. S. (2022). Towards a sociology of catastrophe: The case of COVID-19. In J. M. Ryan (Ed.), *COVID-19: Surviving a pandemic*. Routledge.

World Bank. (2021, October 27). *Global wealth has grown but at the expense of future prosperity* [Press release]. www.worldbank.org/en/news/press-release/2021/10/27/global-wealth-has-grown-but-at-the-expense-of-future-prosperity-world-bank

World Health Organization. (2021, January). *Listings of WHO's response to COVID-19.* Retrieved March 25, 2023, from www.who.int/news/item/29-06-2020-covid timeline

World Health Organization. (2022, December). *Coronavirus disease (COVID-19) weekly epidemiological updates and monthly operational updates.* Retrieved December 24, 2022, from www.who.int/emergencies/diseases/novel-coronavirus-2019/situation-reports

2
IDENTIFYING INTERNATIONAL ESSAYS ABOUT COVID-19

Suresh Nanwani and William Loxley

Study – aims and designs

The book analysis is designed to look at the COVID-19 experience over the years of the pandemic to see how it affected global development prospects. Researchers study this event to assess near and medium-term effects on humanity. Although many assumed COVID-19 would peter out by the end of 2022, there still might be a need for regular boosters or an annual flu-type vaccination, or a one-time vaccination like what is done for polio (Murray, 2022). This assumption was based on current trends in late 2022, as some have expressed hope that booster shots may improve the situation from the past two years.

To accomplish this book project, the editors set up the framework found in Chapter 1 to identify key themes facing society as a result of COVID-19. From this framework, they derived a theoretical underpinning (Briggs et al., 2021). The theory suggests societies can be broken into constituent parts that support specific social, political, economic, and cultural functions. These specific functions operate to keep in equilibrium specific functions of daily life, such as acquiring life skills and gaining awareness.

These key activities operate to maintain overall equilibrium when external and internal forces impinge on particular behaviors in daily life. The sectors interact with one another to support rebalancing forces of change to keep communities, societies, and nations strong. Education and training also serve as conduits to ensure knowledge and information are spread evenly across society. The editors used a case study approach to verify findings and then collected insights about upward social and occupational mobility, public policy strategies, and cultural advances using ICT.

DOI: 10.4324/9781032690278-3
This Chapter has been made available Under a CC-BY-NC-ND license.

The editors collected essays from a select group of professionals that tells various facets of the COVID-19 phenomenon across societal activities. Reporting via these individual essays, they generated an in-depth understanding of the COVID-19 experience in real-life events that transpired from 2020 to 2022. Based on key themes obtained from a sociological perspective, the editors explained and discussed some pertinent and defining issues raised within defined topics that could alter society. The editors hope these insights would satisfy readers' curiosity toward a more informed understanding and provide answers to how COVID-19 has impacted humanity.

The 27 individuals, with ages ranging from the early 20s to the late 70s and a good gender balance, were selected from across the globe. Their essays provide wide-ranging and provoking thoughts that offer a smorgasbord of ideas on a variety of topics for readers to feed, ruminate, and reflect upon.

The development paradigm introduced in Chapter 1 is designed to cover all manner of issues and events that COVID-19 might have impacted in societies around the world (Mattani, 2020; Rutherford, 2019; Babones & Chase-Dunn, 2012). Therefore, the framework incorporates key ideas that appear in the essays and places them in the context of strategies and tactics needed to address COVID-19. This is a good way to keep track of the destructive potential that COVID-19 unleashed on the world, as it affected different regions of the globe and subpopulations within these regions. Case study essays can reveal fault lines and societal weaknesses. As such, the aims of the study are as follows:

1. Identify and understand how people and societies everywhere reacted to and were affected by the COVID-19 pandemic; and
2. Assess the extent to which the pandemic changed global activity, especially upward mobility, work-family, public policy, and technical innovation on cultural values.

The world as a laboratory

The COVID-19 pandemic is a worldwide phenomenon affecting large numbers of people and their outlooks on life. These experiences are the material of the study. Figure 2.1 shows areas where the pandemic has infected and caused deaths in certain regions and countries more than others. Europe, the United States, Brazil, India, Korea, Russia, and Indonesia – all relatively populous nations – reported high numbers of infections and deaths.

These areas have one thing in common: They are reasonably well-advanced or middle-income countries, and they possess health-tracking systems. India and Russia may be exceptions as WHO projected that they might have significantly underreported deaths and infections (i.e., Russia may have had three

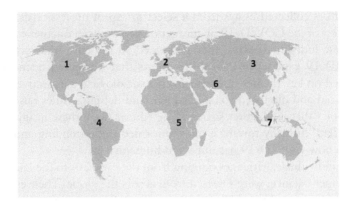

FIGURE 2.1 Map of the world by regions.

times more deaths, and India almost ten times more, though these differences are likely artifacts of system interference from outside the medical field).

Likewise, most of these high population nations have a complex occupational structure in which some jobs were more exposed to the pandemic, especially in urban centers. Population density can distort views about the severity of the pandemic on economic complexity. Hospital workers are a good example, along with people working in close personal care in countries with large-scale public health systems that have many kinds of medical staff.

According to WHO (2022), COVID-19 has caused nations, institutions, and individuals to suffer in varying degrees. Many human activities were affected disproportionately compared to jobs easily converted to WFH. For example, emergency hospital healthcare workers could not stay home. As such, they were severely hit with exhaustion, illness, and extremely difficult working conditions with little compensation. Restaurants and small hotels, making up a vast percent of small businesses around the world, closed their doors, laid off workers, and suffered great economic hardship. By the end of 2022, WHO stated that most businesses were negatively impacted, with USD 17 trillion in lost GDP and emergency public spending on COVID-19. All these conditions altered everyday living and impacted societies around the world. The juxtaposition of public health versus the economy has prompted an ongoing debate on how to handle quarantines the next time there is a pandemic.

Worldwide deaths and infections

Figure 2.1 shows that North America, Europe, Latin and Central America and the Caribbean, and South Asia and Middle East (marked as Regions 1, 2, 4, 6, respectively) suffered the highest infection and death rates from COVID-19 by late 2022. This is in comparison to Region 3 (North Asia), Region 5 (the

African continent), and Region 7 (Southeast Asia together with Australia, New Zealand, and the South Pacific), whose infection rates were less severe.

By the end of 2022, deaths amounted to over 6.7 million and were mainly in the United States (1,100,000), Europe (1,000,000), Brazil and Mexico (1,500,000), and India (1,500,000). Russia and Indonesia each had about 275,000 deaths. Altogether, these countries accounted for 70 percent of all deaths from COVID-19. On average, these countries had large urban populations, well-defined public health systems, and fairly modern data tracking systems. In May 2023, the head of WHO, Dr. Tedros Ghebreyesus, said at least seven million died and that the true figure was likely closer to 20 million deaths – nearly three times the official estimate. He warned that the virus remained a significant threat.

In addition, out of more than 660 million infected people around the world, the following were found in these regions: United States (100 million); Europe (200 million); India (45 million); Latin America – Brazil, Mexico, and Argentina (60 million); and three other countries, namely, Indonesia, South Korea, and Russia (60 million total). While remaining other countries reported over 200 million infections, it is hard to tell exactly how accurate their reported results were.

The significance of highlighting the clusters of infections and deaths in a handful of relatively more modern countries suggests there is some sort of bias taking place. Clearly, these countries were able to report deaths more accurately than less developed nations. Some nations undertook severe lockdowns (e.g., China), while some imposed border isolation (e.g., New Zealand and Australia), even preventing their citizens from returning. Others took a *laissez-faire* approach and let the pandemic run its course. In a few instances, nations faced with dire situations they could not control just downplayed its severity. These explain why WHO estimates the actual infection and death rates are three times higher than what is reported. In contradistinction, there was either little concern over Covid-related deaths or simply an inability to manage the health crisis in many other parts of the world. This anomaly needs to be investigated further as the implications of such patterns for understanding the pandemic can misrepresent global effects on the world.

According to WHO (2022), proportions similar to infection rates arise for death rates across regions. The aforementioned countries also show that infections disproportionately affected six out of ten deaths among poorer than wealthy individuals. Nearly four out of five deaths are the elderly, and urban more than rural, although this is likely due to poor record-keeping in rural areas, along with lower population density limiting contagion. In all cases, COVID-19 set back social progress, especially in public health, schooling, job creation, and as importantly, mental and emotional well-being.

Urban communities in the United States, EU, and the United Kingdom have integrated community living estates for the elderly. COVID-19 became rampant in these centers, killing many elderly as infections spread into the communities unchecked. At the same time, many people with computer connections working from home moved to rural settings to work remotely. They were less likely to be unemployed than, for instance, a personal care assistant or worker, as tourist hotels, airlines, and public event venues closed. Single parents with kids also faced immense challenges – either going to work to earn for their families or staying home with their children. Pandemic effects varied across social classes, communities, and genders. In short, researchers need to keep an eye out for personal and regional bias in survey data.

It is in this context that this book's study was formulated and a case study approach was devised to investigate the impacts of COVID-19 (Mathewman, 2022). As essays on topics relevant to COVID-19 disruptions tell us something useful about their effects on daily life, these should be read in the context of each essay. References are made in the essays to other essays or matters covered in the book, serving as connections for its overall theme, which permeates various sectors affecting our lives.

Structure and organization of the study

Qualitative social survey research covers many domains (Nicola, 2020). Surveys can primarily employ evidence-based case studies in the form of essays. Descriptive analysis starts with identifying whether the pandemic has impacted human behavior. It is an exploratory approach as a prelude to later in-depth analysis and judgment. For example, the intrinsic case studies written by the two editors as practitioner-scholars demonstrate their professional interest in the book's topic. Participant observation in this sense helps readers understand how people changed in response to the pandemic. This, along with essay content and contributors' survey, allows thematic analysis using qualitative data, identifying, reporting, and linking patterns across major segments of society. Though this study alone cannot scientifically generate findings for a wider population, it allows us to garner a wide array of information, which permits researchers to develop hypotheses for further study.

The traditional essay format is helpful in highlighting why and how events transpire. Essays by definition inform, persuade, explain, or entertain. They present coherent arguments in response to questions about COVID-19 and persuade readers that these positions are credible (i.e., believable and reasonable). How does COVID-19 affect urban minorities in big cities? Can overreliance on time spent on online gaming reduce student learning? Is saving the elderly worth the cost of shutting down society? These essays show the insidious impact of the pandemic on humanity (Kjaerum et al., 2021). Then, too, the essay format is conducive to analyzing COVID-19 impacts.

Essay writing provides information and explanation that promotes critical thinking, where reflection reaches a conclusion. As such, each essay presents a topic, a body containing analysis and arguments, and an ending. The essay is a useful format to explore the impacts of COVID-19 on the world and reveal subtle impacts of COVID-19 on daily life, such as how to cope with social isolation and keep the wheels of government and businesses turning. The essay format fits into this project's aim by allowing experience to shape themes, topics, and issues. The essays are grouped into broad themes of the four key sectors of society, polity, economy, and culture.

This analysis is based on selecting a group of experts from around the world to write narratives about what the COVID-19 pandemic has taught them and about the lessons society should take away from it to limit downtime in the future (Creswell & Poth, 2017). The network of essayists is linked to the book's editors, and most are unknown to the others. The writers are professionals and probing thinkers from many walks of life in many fields of expertise, who experienced and thought about how the pandemic has and will impact the world. Young professionals are included to capture the next generation's perspective on post-COVID-19 career choice and development.

We also encouraged the experts to view essays in a wider sense by considering other formats – magazine article, interview, blog, letter, or pamphlet – to explain how COVID-19 accelerated changes from their perspectives. In essence, the essays provide fresh insights and possible solutions that will alter future generations.

Essayists – identification, specialization, and regional coverage

The essayists were chosen based on a global perspective, ensuring a balanced gender mix with different geographical and regional coverage and spanning three generations (Susskind et al., 2020). It took about three months (March to May 2022) to gather 27 experts and place them at the center of the research design (Merriam, 1988). We had initially planned to feature only 12 experts, with the intention that each essay would be about 4,000 words. However, we later opted for more experts to widen the group of writers and limited each essay to 2,000–3,000 words.

Identifying and selecting writers was a challenge, especially when there were many who wanted to contribute their views and write but were not able to do so as they had other commitments. The COVID-19 impacts also made it difficult for some who wanted to share their views but could not do so, as it was a painful process for them to write, and they needed to "move on" despite wanting to share their views. The writers were selected from as wide a geographical basis and gender balance as possible. This approach was constantly kept in mind to deliver varying perspectives of the writers.

The communication process among contributors across continents and time zones posed specific issues. For Zoom conferencing, different computer hardware and software devices presented challenges for transmitting digital material. In one case, power outage caused the contributor's files to be damaged and unrecoverable. Fortunately, through her notes and drafts, she was able to reconstruct her essay. Then, too, contributors who agreed to participate were enthusiastic in spite of busy schedules and personal situations arising from the pandemic. One essayist could not complete the essay on time after contracting long-term Covid. Though several essayists experienced insomnia and severe anxiety due to COVID-19, they were able to complete their essays after taking additional time to pace their schedules. Some had health issues unrelated to COVID-19 and were hospitalized. They, too, managed to complete their essays but needed more time to do so. In general, the contributors comprised able professionals dedicated to the project's goal of documenting COVID-19 effects for posterity based on the sociological constructs premised by systems analysis and structural functionalism.

As is usually the case, it was not easy to strike a perfect scenario. We were impressed by the enthusiasm of the writers, all of whom had their own and unique expertise and experience (see Table 2.1). The writers are from a wide spectrum of occupations and activities: a psychologist, an action research living theory professor, a medical doctor, lawyers, a mediator, postgraduate students, young professionals who graduated from their tertiary education, law professors, a technology strategist, an IT specialist, an international development expert, a research scholar, an ombuds person, a business entrepreneur, an organizational health strategist, a chief accountability officer in an international organization, a researcher who runs bluegrass jam sessions, a trade law and policy consultant, an environmental policymaker, and authors.

TABLE 2.1 Writers from the globe – region and number

Region	Number of writers	Female	Male
Australasia	3	2	1
Asia	8	2	6
Middle East	1	1	–
Europe	3	1	2
Africa	3	3	–
North America	5	3	2
Central America and the Caribbean	2	–	2
South America	2	–	2
Total	27	12	15

Identifying international essays about COVID-19   **39**

Approaching the essayists

We approached the essayists with a concept paper that provided the background for this book (Wilson & Dickinson, 2022). We invited them to submit an essay, suggesting some topics and themes but leaving them to include what they want to focus on. We asked them to define their experiences based on their reflections and shift in thinking, which would help them and others provide insights and possible solutions for the present and future generations. We encouraged the writers to send us their provisional essay title, topics and themes, and a brief synopsis of their essay.

On the basis of their submissions, we had discussions through Zoom and other modes to clarify their thoughts and queries. Writers were given four to five months to write their essays, including the initial brief synopsis of the essay. This worked out best as it suited their timelines and schedules. Most writers requested that we review their draft essays midway before the final submission in October 2022. In some cases, we extended the deadline to November 2022 to accommodate writers who encountered varying problems and issues. This proved to be efficacious in enabling them to finalize their essays. We made it clear that our review was intended to guide them and that our suggestions were for their consideration. The revised essays went through the usual editing process before they were finalized.

Schedule for the book project

The schedule for the book project is provided in Table 2.2.

Clustering themes

In organizing the study, the editors set the format and context for the wide expanse of ideas by clustering Covid related topics and themes with points of view found within. Previous analysis in Chapter 1, along with (i) personal observations, (ii) discussions with others in media, and (iii) the essay titles, all confirmed how we identified the four core sectors. Themes representing issues and topics explain how COVID-19 has significantly impacted modern life, especially in large, advanced, and middle-income nations (Haleem & Javaid, 2020).

These impact areas include (i) *mobility* acquired through education, occupation, and income to meet rising aspirations; (ii) *public sector policy* needed to manage and supervise public services that build trust in the common good; (iii) *diversity in the work environment* needed in a changing economy; and (iv) *technology-led digitalization* needed to promote culture through education, health, business, and information flow to improve the human condition.

Theme One concerns individual mobility in lifelong activities that increase choice, confidence, networking, education, jobs, income, talent, and skills

TABLE 2.2 Schedule for the book project

Sequence	Timeline	Feedback from writers	Experiences gained by editors
1 Book concept paper provided to prospective writers	March 2022	Useful tool for them to formulate their thoughts	A concept paper needs to be flexible to allow changes and adapt for both writers and editors.
2 Initial feedback from writers on provisional essay title, topics and themes, and brief synopsis of essay	May 2022	Explaining contents and helping writers focus on their areas through discussion and clarification of their queries	Realization that others' perspectives can be useful to paint a kaleidoscope of diverse views, for example, interviewing local musicians in a subregion to see impacts of COVID-19 on their occupations and their creative flow.
3 Discussion with writers on information they provided through Zoom and other modes	March to May 2022	Some writers had queries on their approach and wished to clarify the "best" mode, for example, specific topics and themes to focus on, and modality to use, such as interviews, reflective thinking, studies, etc.	Given the global coverage, it was important to be receptive to different time zones and use modes that were appropriate to the context, such as Zoom, WhatsApp, Skype, Messenger, WeChat, and emails.
4 Review of draft essays	June to September 2022	Useful to have another set of eyes to view the initial drafts and offer suggestions for them to consider for enhancement	Realization that initial drafts are works in progress and there is always room for enhancement when different perspectives are shared and discussed.
5 Essays submitted by writers, followed by review, editing, and publication	October to November 2022	Communications and adjustments to improve the essay in focusing on specific matters that writers share with a wider audience or to understand regional and subregional perspectives	Merging the varied essays with golden threads and ensuring connectivity that COVID-19 impacts are best narrated through the expertise brought by the writers.
6 Survey responses	November to December 2022	A few writers sought clarifications. Some gave additional insights.	Merging and extrapolating the responses received.

(Ballantine & Hammack, 2009). It affects seniors, youth, families, single parents, caregivers, workers, and the elderly.

Theme Two is about providing public sector services that effectively support the public good. The process of command and control in policymaking provides reliable information and efficient distribution of services. Science-based findings are especially important in guiding public health policy in times of pandemics (i.e., equipment, monitoring, vaccination, hospital and staff protection, and institutional financial support). All these activities generate faith in society and the public good.

Theme Three relates to business reactions to the pandemic that support employees and working conditions during lockdowns when workers are laid off or work from home. Issues include job-home balance, green living, WFH, office-factory environment, wages and benefits, etc. In all cases, equanimity of access for all is a priority to maintain job satisfaction, mental health, and strong business success through a diversity of choice.

Theme Four is about digital innovation used to improve information flow and cultural exchange in a dynamic society where human development forms values and behaviors, confidence, and personal responsibility, leading to creativity especially in the digital space that gives meaning to society.

Essays falling into these four themes and their titles provide the topics under consideration. Each essay can be considered a case study that describes situations and yields insights about how COVID-19 changed societies around the world.

Survey responses from contributors

As the project progressed, the editors requested contributors to respond to a survey on the COVID-19 impact on their societies (Guttenlog & Struening, 1975). The survey form sent to all 27 essay writers in November 2022 contained a matrix of 16 key life activities taken from the essays, including processes such as social mobility, equality, well-being, personal connectivity, work-family, big data application, and digital entertainment. Alongside each activity were four columns covering the four themes of the book: social (social relations), political (public policy in crisis management), economic (workforce and workplace arrangements), and cultural (technology impact on the internet and social media).

The respondents were provided with the matrix (see Table 2.3) and asked to rate each of the 16 activities for each of the four themes based on a rating impact according to very large (5), large (4), moderate (3), small (2), and negligible (1). From this matrix, it was possible to tabulate 27 subjective assessments to answer two questions: (1) the most and least affected activities by COVID-19 and (2) the most and least affected themes by COVID-19.

TABLE 2.3 Matrix for rating the pandemic on issues and themes

Life activities	Social relations	Public policy in crisis management	Workforce and workplace arrangements	Technology impact on the internet and social media
Public health				
Personal connectivity				
Social mobility and career				
Sense of societal equality				
Well-being				
Emotional IQ				
Community and national resilience				
Work-family stress				
Aging				
Big data collection				
Research and development				
Culture identity				
Childcare				
Self-coaching				
Entertainment, arts, and culture				
Personal health				

The survey carried out in November 2022 was intended to compare contributor impressions to the overall sense of reaction from reading the essays. Ranking the severity of COVID-19 on activities and sectors of society also adds an extra dimension to the research process, which allows hypothesizing about why some activities affected each theme more than others.

Essays

The four key themes identified in the study contain topics and issues supporting each theme. The topics cover issues related to how COVID-19 impacted the theme and changed lives. While many essays focus primarily on a single theme, one essay covers all of them. Therefore, we begin the parade of essays with "Awakening from a paradise lost: Experiences and lesson from the pandemic," written by an investment banker and philanthropist living in the island nation of the Bahamas in the Caribbean. Owen Bethel covers social, political, economic, and cultural aspects of the pandemic, showing the interconnections between themes and impacts that COVID-19 affected society across the board

(Haleem & Javaid, 2020). His essay provides the presentational format followed by other essays in this book.

Reference list

Babones, S., & Chase-Dunn, C. (2012). *Routledge handbook of world system analysis*. Routledge.
Ballantine, J. H., & Hammack, F. M. (2009). *The sociology of education: A systematic analysis*. Prentice Hall.
Briggs, D., Ellis, A., Lloyd, A., & Telford, L. (2021). *Researching the Covid-19 pandemic: A critical blueprint for the social sciences*. Policy Press.
Creswell, J., & Poth, C. (2017). *Qualitative inquiry and research design*. Sage Publishing.
Guttenlog, M., & Struening, E. (1975). *Handbook of evaluation research*. Sage Publishing.
Haleem, A., & Javaid, M. (2020). Effects of COVID-19 pandemic in daily life. *Curriculum of Medical Research Practitioners, 10*(2), 78–79.
Kjaerum, M., Davis, M., & Lyons, A. (Eds.). (2021). *COVID-19 and human rights*. Routledge.
Mathewman, S. (Ed.). (2022). *A research agenda for COVID-19 and society*. Elgar Research Agendas.
Mattani, M. A. (2020). *Cultural systems analysis*. Springer Publications.
Merriam, S. B. (1988). *Case studies in education: A qualitative approach*. Jossey-Bass.
Murray, C. J. L. (2022). COVID-19 will continue but the end of the pandemic is near. *The Lancet, 399*(10323), 417–419. https://doi.org/10.1016/S0140-6736(22)00100-3
Nicola, M., Alsafi, Z., Sohrabi, C., Kerwan, A., Al-Jabir, A., Iosifidis, C., Agha, M., & Agha, R. (2020). The socio-economic implications of the coronavirus pandemic (COVID-19): A review. *International Journal of Surgery, 78*, 185–193. https://doi.org/10.1016/j.ijsu.2020.04.018
Rutherford, A. (2019). *The systems thinker*. Verband Deutscher Zeitschriftenerleger.
Susskind, D., Manyika, J., Saldanha, J., Burrow, S., Rebelo, S., & Bremmer, I. (2020, June). *Life post – Covid-19*. International Monetary Fund. www.imf.org/Publications/fandd/issues/2020/06/how-will-the-world-be-different-after-COVID-19
Wilson, L., & Dickinson, E. (2022). *Respondent centered surveys: Stop, listen, and then design*. Amazon Books.
World Health Organization. (2022, June 29). *WHO situation reports*. www.who.int/emergencies/diseases/novel-coronavirus-2019/situation-reports

3
AWAKENING FROM A PARADISE LOST

Experiences and lessons of the pandemic

Owen Bethel

Introduction

On reflection, COVID-19 has probably been the most transformative global event so far in the 21st century. The "innocence" and routine practices of living in a pre-pandemic world were suddenly shattered and thrown into unprecedented chaos as we attempted to not only identify and understand the source and course of the virus but also simultaneously react to minimize the impact and consequences of a repetitive frontal blow to the face from an invisible opponent. The reality and confusion of the situation resulted in each person being cast into the role of a potential opponent, with isolationist practices becoming the norm at global, national, communal, and familial levels. Fear of the unknown became intertwined with distrust of the known.

The phrase "No man is an island" became the antithesis of actions taken by leaders and citizens alike. Ironically, those leaders and citizens of actual islands in the ecosystem felt the brunt of isolationist actions, since their paradises, the places where "God" and weary souls vacation, came to an abrupt halt in economic activity as borders closed, air and cruise travel suspended indefinitely, and the lifeblood of tourism stopped flowing. For some on those islands in the hurricane belt, the thought of an approaching Armageddon, the prophetic End or Omega, became prevalent: This pandemic event was truly a sign of man's fall from grace and the realization of Paradise Lost.

Precursor to the pandemic

On September 1, 2019, Hurricane Dorian, with category 5 intensity, struck the northern Bahama Islands of Abaco and Grand Bahama, resulting in devastation

DOI: 10.4324/9781032690278-4

This Chapter has been made available Under a CC-BY-NC-ND license.

estimated to be approximately USD 7 billion. There is still no official count of the loss of human life from the tragic event, which was considered to be the worst natural disaster in the recorded history of the Bahamas. While the archipelagic formation of the country allowed life and commercial business to proceed as normal for the most part on other islands of the country, it was inevitable that the shock of the event and the extent of devastation on the two family islands would take an emotional and economic toll on the population and country at large.

Overwhelmed by the degree of destruction left in the path of the hurricane, the government reached out and obtained the support and assistance of the United States, Great Britain, fellow countries in the Caribbean, and a number of non-governmental organizations (NGOs). Not having experienced a hurricane of this magnitude before, the government required supplementary human and financial resources to coordinate first responders and rescue missions; all involved parties were put to the test. This event triggered partnerships among the public and private sectors, civil society and international institutions alike, but still fell significantly short of what was necessary.

Was this the voice of John crying in the wilderness: "Prepare ye the way of Armageddon!"? Was it a harbinger of what was to come, or a test of the resilience of the people in the face of doom and destruction, and faith and hope for a better tomorrow?

Approximately six months later, with still a significant amount of work to be done in the aftermath of Hurricane Dorian, we would experience the reality of that "better tomorrow."

What manner of virus is this?

"The mind is its own place, and in itself can make a heaven of hell, a hell of heaven."

In February 2020, I traveled to Weston, Florida, for my annual health and physical assessment at the Cleveland Clinic. I took the opportunity to enjoy the delights of the Wine & Food Festival on South Beach, and the comedy of Shawn Winans at the Seminole Hard Rock Hotel & Casino in Hollywood with a dear friend. Needless to say, as both venues were packed to capacity and enjoyment, friendly relaxing social interaction were the priorities of the day. Little did we know that this would be the last social interaction of this kind we would partake in for the next two years. The sentence of banishment to the rock of Paradise was the invisible gavel in the judge's hand dancing the tango in the air, with the invisible enemy soon to be called "coronavirus." The mindset of fear that we had unknowingly contracted the virus caused internal panic and paranoia. We closely monitored each recognizable symptom we experienced and surmised the worst as evidence of carrying the virus.

March 2020

In March 2020, the drum roll of concern across the globe about the rapidly spreading virus grew louder and stronger. With it grew the doubts and mistrust of governments and citizens regarding both the source of the virus and methods of transmission. Global geopolitics and medical/scientific expertise fought for recognition as the voice of knowledge and wisdom on the expanding threat. Ignorance and suspicion of the facts and truths expounded by politicians and scientists only exacerbated the problem. These doubts and fears were not lost to the population of the small island states, who felt they were the victims of shenanigans of developed countries and consequently forced into isolation and economic chaos.

Conspiracy theories

Conspiracy theories evolved and spread rapidly by means of social media. This also had a significant impact and influence on the outlook and response of the youth. The older generation was swift to associate the unexplained occurrence as an act of an angry deity, as in John Milton's epic poem, *Paradise Lost* – a consequence of man's disobedience. The youth, however, were more inclined to adopt the conspiracy theory that the virus was man-made for economic gain and that, as young people in their perennial fortitude and privilege, they would be naturally immune to the virus.

This was reminiscent of the HIV/AIDS crisis, perceived by some as a judgment call on a targeted group of people due to their disobedience. This belief resulted in both depression and bouts of rebellion as the youth tried to maintain active social lives and a level of caution and varied levels of observance of the changing legislative and regulatory landscape. This is to help them cope with the effects of the virus, which effectively limited their freedom of movement.

Total lockdown in the island nation

By April 2020, the Islands of The Bahamas were in total lockdown and isolation from the outside world. International flights were cancelled and even interisland traffic was halted, except in the case of emergencies.

The dictate of international institutions

The Government of The Bahamas, like all other governments around the world, was not prepared for this type of unprecedented phenomenon or its impact and consequences. As such, it turned to the voice of internationally recognized scientific and medical expertise – the World Health Organization (WHO) and the Pan American Health Organization (PAHO) – for advice and direction on the implementation of best practices to combat the attack of the pandemic on a vulnerable archipelagic nation with largely open borders.

This reliance on established multilateral organizations was enhanced by the establishment of an intergovernmental agency within the Ministry of Health, mandated to produce, monitor, and review data on various aspects of the pandemic, and make recommendations to the government on necessary measures and actions to be implemented. This agency was later replaced by the National COVID-19 Vaccine Committee when the vaccine became available and was introduced and distributed throughout the islands.[1]

These measures were further reinforced by the introduction and enactment of legislation (Emergency Powers [COVID-19 Pandemic] Regulations, 2020), which created a regulatory arm known as the Competent Authority in the Office of the Prime Minister. It gave the Office or "first among equals" practically sweeping powers to take whatever action was deemed necessary to minimize and eliminate the consequences of the pandemic.

This introduced what appeared to be draconian measures in the form of Emergency Powers Orders as a counter-offensive to the onslaught of the pandemic and the exponentially increasing number of infections and deaths within the country. Actions by the Competent Authority became more of a measure of political expediency camouflaged under the guise of medical and scientific best practices sanctioned by external forces such as WHO and PAHO. Cookie jar solutions had little regard for the indigenous cultural and economic realities, resulting in instances of criminalizing local routine entrepreneurial activity by street vendors.

Healthcare system under siege

The burden on the healthcare services system was enormous and a paramount factor in the government's decision-making process on which measures to implement. This was often exacerbated by bureaucrats attempting to second guess what their political masters wanted to hear and believe. Under normal circumstances, the healthcare system is already strained and in need of structural, technological, and professional improvement. Placing the additional burden of the pandemic's demands on an already challenged healthcare system was a path to disaster. It developed a scenario where decisions of life and death would be left in the hands of doctors and nurses.

The ensuing pandemonium of the pandemic conjured up images of the allegorical capital of Hell given the same name, "Pandemonium," in Milton's epic poem. The isolationist measures, both imposed by external forces and mandated by the Orders of the Competent Authority, caused economic, social, scholastic, cultural, and familial hardships. These resulted in business closures, job loss, student-related stress and anxiety, and a disruption in cultural and familial norms, whether attending church, funerals, or cultural festivities, the daily routine of street vendors and entrepreneurs, or the breach in customarily going to the beach of a family on the weekend.

Tourism sector desecrated

The measures taken by the government, which included the total lockdown of the country, caused the tourism sector, the lifeblood of the economy, to come to a screeching halt, with the complete closure of the airport, borders, and hotels. This certainly had an impact on the workforce, as many employees were either terminated or furloughed. As the situation became prolonged, even those who were on furlough were subsequently terminated, and several smaller hotels throughout the islands closed permanently. This state of affairs clearly had a ripple effect on the social services system, requiring the government to introduce extended salary compensation allowances to the now unemployed hotel workers.

This scenario was then replicated in other sectors of the economy, as construction and other services stopped operations, ultimately causing several businesses to cease operations permanently. The fear of a potential increase in crime and social unrest mounted, particularly as instances of exemptions to the rules for a privileged few were uncovered, resulting in one instance in the dismissal of a senior member of the government. While the existence of a curfew certainly assisted in minimizing, if not avoiding, the actualization of an increase in crimes, there were still pockets of social protest, which were effectively handled by the police force.

Social fabric unraveled

The impact on the social fabric of the country was significant. The prevalent *laissez-faire* way of life or easygoing island lifestyle of the country was shattered, as we recognized that long-standing practices and traditions could be obliterated with the stroke of a pen or the fear that any form of social gathering could be a possible virus super spreader. Picnics, cookouts, regattas, community fairs and festivals, family islands homecoming celebrations, church services, funerals, and weddings were all subjected to significant limitations or prohibited completely.

My younger daughter's wedding ceremony and celebration in November 2020 was reduced to a guest list of ten immediate family members, clearly a devastating and disappointing blow to a young couple who, over a year in advance, had planned to share this most important day with hundreds of family members and friends.

Funeral services, normally seen as both social gatherings and solemn church rituals, were relegated to graveside services and limited to attendance by five (later increased to ten) family members. This created great stress on families which, in the Bahamian tradition, are generally large and closely knit. Observance of the restrictions in the instance of both funerals and weddings was generally upheld. However, in instances where the rules were flagrantly breached and publicized highly by means of social media, no monetary and criminal penalties or imprisonment were imposed. This ultimately made the rules and Orders appear ineffective and unnecessary.

Education

Possibly the most consequential impact of the pandemic on the country will evidence itself over an extended period in the education system. Already burdened with a national average grade pass level of "D" among high school students, the total closure of schools across the country for almost two years, without an effective and practical strategy for providing ongoing remote and online training, will certainly have lasting consequences on the affected generation of students.

The anxiety and uncertainty caused to students will affect their emotional and mental state. This is compounded by the absence of social interaction and requisite socialization in order to propagate essential qualities of the whole man (body, mind, and spirit).

It could be said that isolation during the pandemic, without focused attention and direction into creative, productive, and analytical thinking, would result in a "dumbing down" of students, as they resorted to spending time secluded in their rooms playing games on the internet or virtual entertainment boxes. This will certainly be the subject of significant study and research over the ensuing years, as attempts are made to measure and analyze the real and long-term effect on the youth of both the pandemic itself and the measures undertaken to combat the spread of the virus.

Is the recent surge in criminal activity of stabbings and killings among the student population a result of pent-up frustration, depression, or the lack of socialization during the extended period of the pandemic, and the consequent inability to develop anger management skills?

Culture and entertainment

Culture defines the unique and idiosyncratic qualities of a country and distinguishes it from another. While adaptation and evolution of cultures may result from the interaction of extraneous forces or integration of new elements into a society due to immigration or adoption of perceived best practices from other cultures, the pandemic has impacted the various means of cultural expression more than the culture itself.[2] The pandemic affected all aspects of human activity more profoundly than any crisis before. Around the globe music festivals were cancelled, and the entertainment industry had to quickly shift their business models by innovating and embracing technology to provide digital experiences for their audiences.

Junkanoo

However, *Junkanoo*, a traditional festival of costume, dance, and music held during the Boxing Day and New Year's Day holidays, and the country's foremost cultural identity and event, has defied the changing laws of expression and crowd control in a viral-prone environment. *Junkanoo* continues to attract and encourage the creative juices of a significant cross-section of the country, rising above class, race, age, and ethnic constructs, despite the extremely high probability that the event will be a super spreader for the virus. Attempts to convert it into a virtual event, a government produced "Spirit of *Junkanoo*" virtual parade, with minimum or no spectator participation, failed miserably.

Political change and perspective

> "*Long is the way and hard, that out of Hell leads up to light.*"

A change of government through the electoral process in September 2021 brought a new outlook on the response to the pandemic. However, we should appreciate that by this time, the entire global view of the virus had evolved, primarily due to the existence of a vaccine and booster shots. As waves of new variants of the virus continued to plague the world, the government used the threat of renewed lockdowns was a weapon in successive governments' arsenal. However, the new government had promoted an electoral campaign based on the promise that it has a better way of dealing with the situation and that restrictions in the country would be eased.

The campaign's objective was to get the economy mobilized and people back to work. However, the strategy for the educational system still lacked substance, even though there were efforts to introduce a hybrid learning system of rotating days in classrooms and other days working remotely from home via the internet. Another challenge presented itself, as it became evident that a significant number of students could not afford laptops and the cost of internet service to take advantage of remote learning. Further, in cases where parents had to go back to work, caring for the children left at home became a challenge. Fortunately, the school summer break afforded some respite from the dilemma. By 2022, all academic institutions have been allowed to revert to face-to-face classroom learning.

It became clear from the new government's approach that greater responsibility for cautious behavior was being placed squarely on the citizens, making them directly accountable for their own decisions and actions in the midst of an ongoing presence of the virus.

Never waste a crisis

Prior to 2020, the government had plans underway to digitally transform the public sector over the next decade. However, the pandemic forced its hand as it required immediate digital adoption. The digitization of key services was expedited: processes for passport, driver's license renewal, and vehicle registration were now partially or entirely digitized.

Despite many job losses, the pandemic brought about creativity and innovation among the entrepreneurial minded. Micro, small, and medium business enterprises invested in digital presences, in a show of agility to the changing landscape. Now, more than ever, the ability to leverage digital tools has become essential to business survival. Many revamped their social media platforms to better connect with their customer bases. Prior to 2020, very few businesses offered delivery, curbside pickup, and electronic payment methods. A good example of a business that met the demand was Kraven, a food and beverage delivery company, which continues to thrive even in the new normal.

Awakening to the path forward

> *"Awake, arise or be forever fall'n."*
> *"Heaven is for thee too high*
> *To know what passes there; be lowly wise.*
> *Think only what concerns thee and thy being;*
> *Dream not of other worlds, what creatures there*
> *Live, in what state, condition, or degree,*
> *Contented that thus far hath been revealed."*

The experience we gained and lessons we learned from the pandemic have awakened us from the nightmare and chaos we encountered along the way. We are wiser and hopefully better prepared to handle a similar crisis, in our response, capacity building, and infrastructure requirements. Lockdowns should only be considered by the government as a last resort if all other medical and scientific measures have been exhausted.

Cautionary measures such as mandatory mask-wearing could and should be the first step to curb the spread of a virus. Limiting or restricting social interaction and economic activity should not be contemplated. Citizens should be made to understand that the responsibility for their own lives and that of their family members is in their hands, and that the government will avoid a paternalistic approach to the ensuing crisis, focusing only on the care of the indigent in the community as a priority.

Economic diversification

Successive governments of the Bahamas have paid lip service to economic diversification, moving away from reliance on the primary economic activity of tourism and investing in initiatives for greater self-reliance on food security. But very little has been done to actually move the country in that direction or encourage entrepreneurial endeavors in diverse sectors, particularly the progressive decline in the offshore financial sector. Both the hurricane and pandemic have exposed the vulnerability of the country to sensitivities in the movement of people, goods, and services across borders and supply chain issues, whether caused by natural or man-made disasters.

While hurricanes may cause significant damage to the infrastructure of the country, the pandemic has longer-term ramifications on the health, welfare, educational, and emotional stability of the nation, factors of which are difficult to measure or quantify. Food supply security is an issue that straddles both crises of hurricane and pandemic. Concerted efforts must be made, in the context of a multisectoral approach, to ensure the country and its people are resilient in the face of such disasters.

Climate change toward a new paradise

The Bahamas has a unique opportunity arising out of the combined hurricane and pandemic crises of revitalization of the country. This can be done by awakening the entrepreneurial spirit of the private sector and harnessing the significant growth potential in the areas of sustainable and renewable energy, the blue and orange economies, the digital economy, sustainable agriculture, and sustainable tourism. The country continues to be a model for other small island developing states in many aspects. Despite historical, cultural, geographical, and economic nuances, the Bahamas is an example of how, despite many developmental challenges, resiliency will safeguard our people from future shocks and chaos.

The Bahamian economy can revive with strong tourism and banking sectors, combined with a commitment to full employment and worker wellness. Important cultural events can return with the appropriate and essential social gatherings, while recognizing and upholding the common values of crime prevention and promotion of a strong public healthcare system. Accordingly, the political system can engender and strengthen a civil society, which encourages a norm where both government and citizens take joint responsibility for human wellness.

Similarly, the social system can reboot classroom and blended learning modules, which again teach skills that broaden work and career opportunities and promote feasible social expectations.

A stable society with strong sectors is only created when its population learns from the mistakes it made in handling the response to the pandemic. When all sectors work in tandem, society can again regain the lost paradise. *Junkanoo*, a landmark symbol of freedom and jubilation created and passed down by our slave ancestors, will once again return in all its splendor as a symbol of our freedom from the captivity and chains of the pandemic, a phoenix rising from the ashes of a paradise lost to a new paradise – indeed a paradise regained.[3]

Notes

1 See Chapter 16, "Change and continuity: COVID-19 and the Philippine legal system" by Antonio G. M. La Viña.
2 See Chapter 31, "Emotional and physical isolation in a Latino community" by Alfred Anduze.
3 I owe a debt of gratitude to my daughter, Natalie, for her youthful wisdom, insight, and contribution to the article, and for providing me with the perspective of the next generation, which will ensure the survival and resiliency of paradise regained.

Reference list

Milton, J. (2003). *Paradise lost* (J. Leonhard, Ed.). Penguin Books.

PART II
Educational opportunity and social mobility

4
INTRODUCTION TO THE SOCIAL SECTOR

William Loxley and Suresh Nanwani

Introduction

Upward mobility into a higher social class is the oil that lubricates the wheel of progress for students, workers, business people, artists, thinkers, innovators, and policy analysts (Brown et al., 2013). Raymond Boudon (1974) suggested that upward mobility demands acquiring education, occupation, income, tech-skills, self-confidence, social awareness, and aspirations for career and family formation. The opportunity to experience choices generates greater knowledge transfer useful for a better future. Wider networking found in inclusive social and professional networks allows many in society to capture the most and best talents available. Besides social networking, upward mobility creates self-esteem and personal confidence in adapting to work, family, community, and cultural awareness, culminating in personal human development.

COVID-19 forced most people into social isolation for long periods, limiting face-to-face expression and only allowing contact on the digital world. This is especially true in the developed world (Breslin, 2021). The situation prior to vaccine availability, along with school, work, and other institutional shutdowns, left many without a daily routine to follow on their personal journey to growth and maturity. Disruption affected the well-being of young people due to fear of an unhappy future, leading to a lost generation of dreams. This loss of confidence ultimately led to a loss of creativity when economic security was jeopardized (Johnson & Salter, 2022). One wonders if the financial return to education is worth the social mobility and mental well-being of students and workers who were furloughed after many years of employment, or to those who were nervous about quality public health and personal wellness.

DOI: 10.4324/9781032690278-6
This Chapter has been made available Under a CC-BY-NC-ND license.

Much of the formal news media throughout 2020–2023 suggested that education and training in society suffered greatly throughout the pandemic (Di Puetro et al., 2020). Upward mobility stalled as education, careers, and income stagnated due to limited opportunities. Other reports suggest depression among youth concerning fears of career prospects and skill development (Kraus & Park, 2014). Emotional well-being, aspirations, and general human development declined for many as individuals and groups felt blocked from achieving their ambitions.

Advanced cognitive skills are deemed necessary for social progress. However, critical and creative thinking were stunted as advanced skills in cognitive reasoning faltered (Burgess & Sirversten, 2021; Breslin, 2023). These in turn spotlighted intergroup inequality and a perceived lack of fairness. Generation Z was inhibited from learning technical skills needed to enter the realm of the metaverse as entrepreneurs. For other age groups, lifelong learning was put on hold, preventing the youth a means to keep abreast of the changing digital landscape.

COVID-19 limited social interaction, preventing social relations from maturing. Among senior citizens, the lockdown accelerated isolation from medical information, ICT skills, and a general attachment to lifelong learning pursuits (Boulton-Lewis & Tam, 2012). In a time of crisis, the elderly were shut off from knowledge channels that offered physical, mental, and emotional well-being. In this period of reflection, many were left to seek solace alone. Finally, the pandemic forced all segments of society to stay ahead of the learning curve, allowing technology to work its magic on economic and cultural growth (Dede & Richards, 2020; Upor, 2023).

New learning practices can be used to revise school curriculum and instruction. John Hattie of New Zealand suggests that modern classrooms do best when they are student-centered. Instructional design is seen as the most effective way forward for students to master advanced thinking skills. While new learning technologies can engage human development, this process was retarded under COVID-19. Because of this, many student cohorts will suffer the consequences (Hattie, 2009; Paunov & Planes-Satorra, 2021).

As nations advance based on preserving and creating knowledge, governments are tasked with transferring human and cultural values supporting the common good. Education becomes a key in transmitting knowledge and values across generations. This is why nations encourage universal education to prepare the youth for the future. As parents, siblings, peers, and mentors are taught learning appreciation, one can only imagine how this process fared under the extended lockdown.

The individual learns sensitivity to contradictions in the thought process of give-and-take based on experience. This progression of mastering thinking skills leads to self-satisfaction, where the self merges with the social system to model human character. The modern world possesses many ideas and methods employed in the economy, culture, polity, and family life (Haidt, 2011). This collective wisdom about fairness, loyalty, trust, and respect can aid the institutions of government, business, community, home, and school to promote the virtues of society that keeps nations on track in good and bad times.

Six essays

Essays in the book fall under four themes about how COVID-19 has impacted humanity. The essays in this chapter fall under Theme One: personal mobility derived from educational opportunity and new experiences, which provide choice and give confidence to those developing their life skills and ambitions. Education is a central institution to the social order because it provides mobility from teaching skills that socialize the young to get jobs and earn income. It also introduces them to the ways of society (Parsons et al., 1969; Strawser, 2022).

The issues found in the essays include looking at youths, senior citizens, rural minorities, and community concerns to observe how COVID-19 allowed different points of view to arise, affecting especially learning opportunities, physical health, and emotional health. Topics include remote learning during quarantine lockdown, social isolation and career guidance, integrated community, virtual education, life goals among the rural poor, and youth and elderly perspectives.

Reference list

Boudon, R. (1974). *Education, opportunity, and social inequality*. John Wiley and Sons.
Boulton-Lewis, G., & Tam, M. (Eds.). (2012). *Active ageing, active learning*. Springer.
Breslin, T. (2021). *Lessons from lockdown: The education legacy of Covid-19*. Routledge.
Breslin, T. (2023). Schooling during lockdown: Experiences, legacies, and implications. In J. M. Ryan (Ed.), *Pandemic pedagogies: Teaching and learning during the COVID-19 pandemic*. Routledge.
Brown, P., Reay, D., & Cinvent, C. (2013). Education and social mobility. *British Journal of Sociology of Education*, 34(5–6), 637–643. www.tandfonline.com/doi/abs/10.1080/01425692.2013.826414?journalCode=cbse20
Burgess, S., & Sirversten, H. (2021). *Schools, skills, and learning: The impact of COVID-19 on education*. VOX EU Multimedia. https://voxeu.org/article/impact-covid-19-education
Dede, C., & Richards, J. (Eds.). (2020). *The 60 year curriculum: New models for lifelong learning in the digital economy*. Google Books.
Di Puetro, G., Biagi, F., Costa, P., Karpinski, Z., & Mazza, J. (2020). *The likely impact of COVID-19 on education*. European Commission.
Haidt, J. (2011). *Why good people are divided by politics and religion*. Pantheon-Knopf.
Hattie, J. (2009). *Visible learning: A synthesis of 800 meta-analyses relating to achievement*. Routledge.

Johnson, E., & Salter, A. (2022). *Playful pedagogy in the pandemic: Pivoting to game-based learning*. Routledge.

Kraus, M., & Park, J. W. (2014). The undervalued self: Social class and self-evaluation. *Frontiers in Psychology*, 5(1404). https://doi.org/10.3389/fpsyg.2014.01404

Parsons, T., Coleman, J., Inkeles, A., & Dreeben, R. (1969). Socialization and schools. In *Harvard education review*. Amazon Books.

Paunov, C., & Planes-Satorra, S. (2021). *Science technology and innovation in the time of COVID-19* (Policy Papers, 99). OECD Science, Technology and Industry. https://doi.org/10.1787/234a00e5-en

Strawser, S. (Ed.). (2022). *Higher education: Implications for teaching and learning during COVID-19*. Lexington Books.

Upor, R. A. (2023). Adapting technology in language teaching and learning in Sub-Saharan Africa: Lessons from the COVID-19 pandemic in Tanzania. In J. M. Ryan (Ed.), *Pandemic pedagogies: Teaching and learning during the COVID-19 pandemic*. Routledge.

5
PERSONAL DEVELOPMENT IN A TIME OF CRISIS

Frédéric Ysewijn

Introduction

The COVID-19 pandemic has undeniably impacted people in asymmetric ways. Thus, I need to preface this essay by informing readers that I know the challenges I faced during this pandemic are in no way comparable to the economic hardship and life-threatening situations some of my cowriters have faced. As such, this essay can best be read from a reflective point of view rather than a depiction of the intensity of this pandemic.

In January 2020, life seemed to go as planned. After four years of dedicated work during my undergraduate degree, I was accepted to continue my academic journey at the London School of Economics (LSE), where I would be reading political economy. Not only has this been my goal for several years, it also instilled within me the truth that when I was willing to work hard, I could achieve my personal goals. Moving forward to March 2020, I realized that excitement could quickly lead to trepidation when my external environment started to change dramatically. Over the next months, it became clearer that a new virus, COVID-19, would have an unprecedented impact on both modern society and our individual modes of being. During the first month of the pandemic, emotions of uncertainty and distress installed themselves into the human consciousness and opened a new unwanted chapter in everyone's lives.

Although individuals and communities have experienced the COVID-19 pandemic in their own ways, I think the feeling of uncertainty around the present and future is the best common denominator during its onset. Common questions were: What will the COVID-19 numbers look like today? When will this all end? For me, the question, "When will I be able to return to university

DOI: 10.4324/9781032690278-7
This Chapter has been made available Under a CC-BY-NC-ND license.

to finish my bachelor's degree?" kept running in my mind in an almost perennial manner.

If there is one occasion I will never forget during those first couple of months, it was when I asked my parents how this would all play out. They answered that they were equally in the dark on how this pandemic would further develop. That was when I knew things were different, as up until that day, both my mother and father were able to provide solutions or advice on how to overcome almost any challenge. This was the moment that my life and that of many others pivoted toward what I call a paradigm shift, where the only certainty was uncertainty.

Collective solitude

Politicians, healthcare experts, think tanks, and civil society all had their own views of how to handle the uncertainty. But when it came to self-quarantine, all that outside noise seemed to fade, until all that was left was collective solitude. One could argue that for the first time since World War II, our society as a whole moved away from maximizing productivity and toward trusting everyone's individual agency to overcome this crisis collectively. In practice, this meant that from one day to the other, society had to reconfigure itself by transposing the real world onto our digital screen. For me and many others my age, it meant that I found my classmates in front of my laptop. Everyone spread out across the globe showed uncomfortable but childish excitement with a smile on the screen, of what soon became a new era of digital education and work environment.

During those first months up until September 2020, I had the luck of being able to continue my schoolwork and summer internship online. Slowly this digital mode of existing became the new normal, but I was never able to alter my inclination toward a more in-person way of doing things. Nevertheless, I must confess that the digital transition and benefits of hybrid work and school models would never have occurred as fast without the pandemic. Suddenly, online teaching became at par with in-person education systems, allowing a much wider range of people with inquisitive eagerness to receive world-class education, often at a more democratized price.

Higher education in London

By September 2020, I had to pack my bags and move to the United Kingdom, where I would stay for the next year to pursue my master's. During my time there, I was given the opportunity to experience a blended learning model of education, which entailed in-person seminars in conjunction with online lectures. The minor price I had to pay for LSE's efforts to keep in-person classes a reality was to take a Covid test every three days for the whole academic year.

During that time, I was living together with two kind and intelligent students from India and China. During our many dinners together, it became apparent that all of us were facing our own challenges and were worried for our loved ones spread out all over the world. Nevertheless, if this pandemic has taught me one lesson, it is that living or staying in contact with friends can undeniably assuage the difficulties of living through hard times.

Around November 2020, uncertainty started to wane. There was hope that we were approaching the tail end of the pandemic, combined with the news that vaccines would be made available soon. This hope, however, was soon diluted due to new variants that brought COVID-19 back into the center of public debate. Additionally, when the first vaccines arrived, it challenged my naive understanding that once there was a vaccine, all our problems would be solved. This, of course, was a simplification bias I had, as I did not take into account the intricate nature of global supply chains to get a global vaccine effort up to speed.

Aside from the medical aspects of the COVID-19 pandemic, this period also let me reflect on what it means to be a global citizen during an age of globalism in retreat. I also thought about how I could develop new and authentic relationships during a time when serendipity has been significantly hampered by the multitude of lockdowns. Additionally, this period also provided modern society with an uncomfortable reality around individual responsibility. This is especially true in Western democracies, where people have, according to some sociologists, become obsessed with individual freedom and self-actualization, at the detriment of collective responsibility in times of crisis.

Individual liberty versus collective responsibility

This tension between individual liberty and collective responsibility could best be perceived during the protests I observed in London around the lockdown measures and vaccination campaigns. I remember how once, when I went in a restaurant while wearing my mask, two individuals shouted at me to take it off and stop being scared. As I did not see the need to enter a verbal altercation, I responded that it was my personal preference to keep the mask on, before wishing them a pleasant evening.

On another occasion, I found myself in the London tube during a heated debate between an elderly man and a teenager who refused to wear his mask when the former requested it. Both experiences taught me how fragile societal stability and respect for one another inherently is, and that external events can quickly lead to forming the key ingredients for civil unrest.

During my year in London, I had to start looking for jobs. I had to go from a learning to earning phase. Southeast Asia has always fascinated me, and more specifically, I started to have an inquisitive interest in working for the Asian Development Bank, which has its headquarters in Manila, Philippines.

Nevertheless, this idea of moving to Southeast Asia became like a pipe dream due to travel restrictions and ongoing lockdowns. I decided that it was best to find a job in my home country, Belgium, and started applying for several positions. After a few applications, I finally got the opportunity to interview for a firm in Brussels, which focused on topics and key issues I found extremely interesting, such as international trade, EU policymaking, and financial regulation. Once I started the job at the end of 2021, a sense of calmness came back to me. I knew I had gone through the storm of the pandemic and that things were finally starting to look better.

A new year and another problem

Unfortunately, my burgeoning optimism was quickly challenged during the first quarter of 2022. Aside from atrocious human suffering, the recent Ukraine conflict caused a significant energy crisis that concurrently placed downward pressure on the macroeconomic environment. As uncertainty is creeping in once again on society, it is imperative to retain what we learned from the pandemic – that our built-up resilience is pivotal to not succumb to the distress of these events.

In other words, if we give over to our emotions, we can easily fall into a fatalistic cycle that does not provide us with any purpose. In making this assessment, I do not support falling into the unnecessary type of Panglossian, or excessive, optimism, but advocate for trying to find some positivism within the realm of reality. Hence, I believe we need to find solace in the fact that the only certainty in today's day and age is uncertainty.

Summing up

COVID-19 taught us in a very clear manner that despite all the uncertainty, one needs to continue navigating daily life. It is imperative especially for young adults to accept the reality around us and accept that instability and uncertainty remain a constant. Although this assessment sounds daunting, the COVID-19 pandemic has shown the inherent value of keeping a positive attitude and appreciating friendships that provide holistic support networks.

Additionally, the general assessment made in this essay requires us to promote a mindset of agility and flexibility embedded within the spirit of pragmatic observation. At the same time, the pandemic also caused us to realize the importance of work-life balance and the inherent benefits of democratizing online working and learning models. In short, and despite the devastating impact this period has had on many lives, I urge everyone to reflect and gain insight from our recent past to make a better tomorrow.

6
TAILS WE GO, HEADS WE STAY

Ariel Segal

Introduction

Life-altering decisions are tough. When every path laid before you is shrouded in uncertainty, it can take days, even weeks, to come to a decision. Sometimes all you have is a handful of hours. As COVID-19 spread rapidly across Spain, I watched my friends pack their things while others hunkered down for the long run. I was caught in a dilemma. My partner (Norwegian) and I (Israeli) had to decide whether we would stay in Spain or go back to our home countries in a mere 6-hour window – before Madrid airport closed and the last flights took off. While I was leaning toward the decision to leave, he was adamant on staying in our flat with our roommates. Time was running out. It was dramatic, to say the least; the future had never been so uncertain for any of us. Rather than spending what could be our last few hours together locked in discussion, we flipped a coin. A life-altering decision made in a split second – whatever the coin would say, we would follow.

When we first heard about COVID-19 spreading across China, I was one of those who did not give it much concern. As International Relations students, we had various discussions about misinformation, globalization implications, and governmental services concerning the disease. Yet it still felt distant. There was growing panic and uncertainty in the air among the international students of the IE community. Those whose countries were already impacted by the disease were the first to grasp the severity of the approaching pandemic. Students and staff from China, Singapore, Japan, France, and Italy were among those who saw COVID-19 for what it was and what it could become. Some served as channels of information about the disease to their close ones back home, as not all countries were highly informed.

DOI: 10.4324/9781032690278-8
This Chapter has been made available Under a CC-BY-NC-ND license.

My thesis supervisor was one of these people, and she shared with me how she was helping her family in Kenya understand the approaching threat. Reading the concerning news abroad, some, like myself, were hoping that the media was exaggerating. We had our undergrad thesis and last year of university to focus on. Any time spent worrying about the disease meant time taken away from our bachelor degree's most pivotal moments. Retrospectively, it seems the world was split into three types of people: those few who saw the disease for what it was and could become; those who did not yet hear about it; and those, like myself, who underestimated it.

Spanish government's and media's response

The Spanish government's and media's initial response painted a picture of the disease as an external threat imported by tourists, a terror-wreaking virus brought by travelers. The first reported case of SARS-CoV-2 in Spanish territory was a German citizen vacationing in the Canary Islands. This supported the comfortable belief that the disease remained external, confined to countries like Italy, rather than a manifesting Spanish public health crisis. On February 26, the first Spanish citizen from Sevilla tested positive, without traveling abroad, followed quickly by the first recorded death on March 1 in Valencia. This signaled the beginning of the Spanish coronavirus nightmare, the swiftest spread of the virus in Europe per capita. According to research done by the Center for Systems Science and Engineering (CSSE) at Johns Hopkins University (JHU), as of September 9, 2020, the toll of the virus in Spain reached ninth place in terms of deaths and ninth in the world in terms of the number of confirmed cases per capita (Heanoy et al., 2021).

Uneasiness spread quickly through Madrid. In the hives of the education system, a bottle of hand sanitizer was passed around at the beginning of each class. That was as far as it went at first. This was followed by the optional mask in February. By mid-March, most classes were taught online. Between February and March, professors at my university expressed concern for their health.

The fear of the unknown virus combined with exposure to an international community that probably traveled more than it should have put many on edge. IE University was better equipped than most educational institutions, as resources for online classes were already up and running for students taking semesters abroad. Switching to online classes entirely was an easy choice given the circumstances. Yet that decision came too late, and the IE community became one of the coronavirus playgrounds in Madrid and Segovia (Royo, 2020).

Drawing on theory

When I was approached about sharing my experience of the epidemic for this book, I used Erik Erikson's stages of psychological development to help understand and convey my own experience. In his book *Childhood and society* (1993), Erikson discusses various conflicts that shape individuals' personalities at eight different stages of life. Stage one: infancy, learning to trust or mistrust your surroundings. Stage two: early childhood, learning autonomy while dealing with feelings of shame and doubt. Stage three: play age, learning to plan and execute while facing guilt in attempts to gain independence. Stage four: school age, experiencing the pleasure of successful problem solving or the disappointment of failure when others around you succeed.

Stage five: adolescence, figuring out identity and developing a lasting sense of self. Stage six: early adulthood, opening ourselves to intimacy and becoming a part of the wider world. Stage seven: adulthood, discovering a sense of purpose and ways to contribute to the world. Stage eight: old age, reflecting on a lifetime of decisions with a sense of satisfaction or failure. If left uncorrected, problems arising in earlier stages might have a snowball effect, leading to a greater psychological imbalance in later stages.

When I used Erikson's theory to understand my experience and the experiences of those around me, it became easier to imagine the many ways people's lives were affected by the pandemic. Suddenly, parents had the option to spend more time at home, offering a closer supportive system to children who were learning to trust their families as well as themselves. However, as we move past childhood, relationships outside our own family begin to play a larger role in shaping our virtues.

Early adulthood, my own generation, is when young adults explore their identity, build relationships with strangers, and develop long-term social commitments. When one successfully grows in this stage, it can result in a greater understanding of oneself through meaningful relationships. However, when the world shut down and intimacy was difficult to find, the result of isolation, particularly in this stage of psychological development, is loneliness, social anxiety, confusion, and depression.

Decision to depart from Spain

By late March 2020, airports around the world were closing. Military-led lockdowns were enforced as the virus reached most corners of the world. My friends and classmates disappeared to their home countries without warning. At the time, I lived in central Madrid with my partner from Norway and two close friends from Venezuela and Switzerland. All of us were trying to figure out how to complete our last university year and bachelor thesis online. Our ignorance started to fade as reality knocked on our doorstep. The city became eerily quiet. My partner laughed uncomfortably as speaker-carrying drones patrolled the street outside our windows, urging the population to stay indoors. Funeral cars and ambulances rush by ever so often. Looking back, it was quite like the start to some dystopian Hollywood film.

As my family expressed their concerns about the worsening condition of Madrid, my partner and I began contemplating our options. It was difficult, we couldn't come to an agreement, and the last outgoing flights were only a few hours away. On March 20, we simply tossed a coin, letting fate decide what we do next. Heads – we remain coolheaded with our friends and face the uncertainty together. Tails – we tuck tails and fly home to our own countries and close families.

From Madrid to Tel Aviv

I won and naively said, "I will see you in a month or so" to my partner and friends. A total of five hours passed between the fate-deciding coin toss and when I sat down on the last available seat from Madrid to Tel Aviv. The flight was tense, like nothing I had ever experienced before. Passengers wore masks of varying designs and complexities, gloves, hand sanitizers, and some even wore full hazmat-like suits.

When I arrived in Israel, I realized that their initial reaction to the disease had been very different to what I experienced in Madrid. Each recorded COVID-19 case was published in the media with an assigned number and a list of locations they had recently visited. This enabled some citizens to track their relativeness to confirmed patients.

Additionally, the government utilized phone surveillance systems to inform citizens via text message if they had been near a confirmed patient and needed to go into isolation. Although it raised uncomfortable questions about privacy, it quickly became the most popular game in Israel. "Have you been close to Patient X?" While I was isolated in my room for two weeks, my phone was constantly abuzz with Covid alerts from the government (Leshem et al., 2020).

Back in Israel

Israel's COVID-19 response is an interesting case study (Coronavirus Research Centre, n.d.). At the time of the outbreak, the Israeli healthcare system faced a chronic shortage of healthcare resources. It would also be an understatement to say the country was amid a political crisis. However, when the epidemic was first reported, Israel's Ministry of Health (MoH) implemented a strict strategy of containment. It included travel restrictions, extensive testing, self-reporting quarantine, the aforementioned phone surveillance system, and a budget to support those afflicted and unemployed. All these items were implemented before April 2020, a comparatively swift initial response to the pandemic.

This strategy proved successful in delaying the spread of the virus among the general population, resulting to smoother peak transmission compared to other countries across the world. This enabled the Israeli healthcare system to gain an advantage by implementing medical staff training, shifting resources in the hospitals, and recruiting support from medical military personnel before new waves and variants of the disease swept through the nation.

Nonetheless, there was high public unrest as schools and businesses closed and lockdown regulations were enforced. It seemed like some countries found it easy to stay at home, but that was not the case for Israelis. Many were looking for loopholes in the rules to move around and visit anyone they could – seeking out intimacy, even at a cost, to avoid social isolation and its impacts. The public disproval toward the government, specifically at-the-time Prime Minster Netanyahu, grew, which led to protestors taking to the streets. Compared to Spain, the Spanish population seemed to be more present and understanding of the strict, military-led lockdown, while the Israeli population focused on the criminal offenses by Netanyahu and the unstable government. One could say that the protests against Netanyahu offered an obvious scapegoat for many individuals whose psyche craved connection, intimacy, and purpose.

In December 2020, Israel started a mass vaccination initiative by administering the BNT162b2 vaccine (Goldberg et al., 2021). As Israel was among the first to distribute the vaccine on such a mass level, the public was suspicious. Distrust in the government was high, and the large spending on acquiring the vaccines, along with the feeling of being the world's "lab rats" for the new vaccine, kept many on edge. While many European countries were skeptical, the mass vaccination campaign led to a sharp curtailing of the outbreak. As of July 2022, over 18 million doses have been successfully distributed, fully vaccinating 71 percent of the population. Thus, Israel became a world leader in COVID-19 vaccination.

Adapting to life online

Adapting to life online was strange and difficult, particularly for the elder population. I witnessed family relatives sitting around, unsure of what to do, day after day, waiting for the pandemic and lockdowns to end. Stuck in a state of stagnation, they felt a lack of purpose and even failure. Younger friends and family kept themselves relatively active, facing the new challenges the lockdowns created. Life online tested social integration, connectivity, and establishment of one's self-identity. My younger brother would set for himself daily goals with both school and personal projects and found new virtual ways of interacting with his friends.

I was already used to keeping in touch with people online as most of my close friends live in different places around the world. While it was not as real of a challenge as some faced during these times, retrospectively, I could tell that the impossibility of building intimate relationships at the time has left its mark. The uncertainty of when I would see my partner was the hardest aspect. My daily routine changed accordingly, as falling asleep and waking up with my phone became the new norm. Getting ready for the day became a significantly shorter process. Reality and the severity of the epidemic sunk in as weeks and months went by. The people I kept in touch with were struggling to a certain extent with the effects of isolation, each in their own way. Loneliness and boredom were not uncommon. I tried to focus on my thesis research, and like many others, I figured it was a good time to adopt some new hobbies like cooking and exercise.

Completing my thesis

Eventually, I successfully completed my thesis research about *Sustainable water management for peacebuilding and economic development* and had the strange experience of defending it remotely. With all the challenges at the time, I was proud to have worked on a literature gap and cherished the 9.5/10 grade I received from my thesis committee. Shortly after, I started working for a large global company on their corporate marketing team as a project manager. I learned the importance of acquiring work experience, especially during the present pandemic.

I never thought I would work only remotely with an American team I had never met in person. Work, meetings and hangouts, hobbies, friends, and education – the past years have proven that humans can endure living online in the metaverse to a certain extent. I believe that this transition is exponential and that we got a premature taste of what life will look like as younger generations take their hold. To me, the biggest challenge was balancing screen time with social interaction, as the introverted part of me was gradually taking over.

Norway

After almost a year since I left Madrid, painstakingly navigating documents and permits, COVID-19 tests, and a great number of hours in quarantines, I finally got to visit my partner in Norway. Seeing and being with someone you love cannot be compared to talking online. My travel back and forth was the only way we could see each other, as Israel stayed mostly closed to foreigners from February 2020 to May 2022. It was the same with the rest of my friends outside Israel.

Suddenly and dramatically, several aspects of my life that were intimate and part of my identity, things that had taken me years to build, were completely put on hold. It was tough, and although I wonder at the effect all this might have on my future personality, I count myself as one of the lucky ones. Some could not continue their work and studies remotely, some had to close their businesses and lost all financial assurance, and some could not see their loved ones for years. Some could not stay "indoors" because they simply did not have the means, and others lost people they care about.

Returning to the "next normal"

Overall and as of July 2022, things seem to be returning to normal, especially in Israel. All regulations have been lifted, and the coronavirus is rarely mentioned other than as a conversation filler. It is treated like any other influenza, requiring the Israeli population to minimal quarantine if tested positive. Businesses are reopening, travel is returning, and schools and industries are back to full capacity. It also seems that society has progressed as a whole, with workplaces and schools becoming more open to part-time and full-time remote work (Israel's response, n.d.).

Considering the environment and how quickly we made a positive dent in carbon emissions due to decreased daily commutes, as well as the huge strides we have taken in the pharmaceutical industry, it is times like these when it is easy to keep a positive spirit. If identity and intimacy are as important to early adulthood as Erikson suggests, then we will see the true consequences of the pandemic as our generations pass. As we collectively traveled through a dark and uncertain tunnel, I hope we have come out into the light stronger and wiser. I am inspired to see what future challenges the years of living through COVID-19 might have prepared us for.

Reference list

Coronavirus Resource Centre – Israel. (n.d.). *World countries: Israel.* Johns Hopkins University. https://coronavirus.jhu.edu/region/israel

Erikson, E. H. (1993). *Childhood and society* (2nd ed.) W. W. Norton & Company.

Goldberg, Y., Mandel, M., Bar-On, Y. M., Bodenheimer, O., Freedman, L., Haas, E. J., Milo, R., Alroy-Preis, S., Ash, N., & Huppert, A. (2021). Waning immunity after the BNT162b2 vaccine in Israel. *The New England Journal of Medicine, 385*(e85). www.nejm.org/doi/full/10.1056/nejmoa2114228

Heanoy, E. Z., Nadler, E. H., Lorrain, D., & Brown, N. R. (2021). Exploring people's reaction and perceived issues of the COVID-19 pandemic at its onset. *International Journal of Environmental Research and Public Health, 18*(20), 10796. https://doi.org/10.3390/ijerph182010796

Israel's response to COVID-19: Strengths, weaknesses and opportunities. (n.d.). *Near East South Asia center.* https://nesa-center.org/israels-response-to-covid-19-strengths-weaknesses-and-opportunities/

Leshem, E., Afek, A., & Kreiss, Y. (2020). Buying time with COVID-19 outbreak response, Israel. *Emerging Infectious Diseases, 26*(9), 2251–2253. https://doi.org/10.3201/eid2609.201476

Royo, S. (2020). Responding to COVID-19: The case of Spain. *European Policy Analysis, 6*(2), 180–190. https://doi.org/10.1002/epa2.1099

7
ADAPTING TO VIRTUAL EDUCATION DURING THE COVID-19 PANDEMIC IN PERU

Victor Saco

Introduction

The first official case of COVID-19 occurred on March 6, 2020. Previous to that, there was only speculation about the pandemic. In general, people in Peru did not feel it would be severe because news about previous similar pandemics such as SARS never spread to the country. However, on March 16, only ten days later, the first quarantine started, which was strictly extended until July 2020. In many cases, people were forbidden to leave the house throughout the day. Only those with special authorization or emergency reasons could travel the streets for short periods.

There was fear of contagion throughout the population, with a few groups rejecting strict and unfounded measures they felt limited their personal freedoms. There were cases where the police took people to jail just because they walked their dog in front of the house, or where the police used excessive force to break up birthday celebrations within homes. There was even a case where 13 people died in a clandestine party – as the police intervened and blocked the exits, some died in the stampede while trying to escape the police. Still, the little scientific evidence available about the disease and the lack of technical debate on television generated a general acceptance of these quarantine measures, despite restrictions on individual rights and negative economic impacts on the larger population.

My personal recollections, translated into English, tell how the university community met the challenge of COVID-19 quarantine restrictions on the traditional instructional methods influenced by the Spanish Lecture System in place for many decades. I further assess how COVID-19 may permanently impact university study after the pandemic subsides.

DOI: 10.4324/9781032690278-9

This Chapter has been made available Under a CC-BY-NC-ND license.

Before discussing the educational response to the pandemic, I need to share a few points about COVID-19's impact on Peru. After the country's initial reaction in 2020, Peru went through intermittent quarantines for the next two years. The beginning of 2021 was particularly bad, when the new Delta variant struck Peru badly. In 2022, society was able to manage the pandemic better. There was a return to pre-pandemic behaviors, such as face-to-face instruction throughout the education system, but with safety precautions of social distancing.

Peru, a land famous for mountain wonders, including Machu Picchu, and jungle ecosystems with many plants and animals, has a population of around 33 million inhabitants. COVID-19 caused 4 million infections and 200,000 deaths in the country, over half of them in the capital, Lima. Higher education in Peru is tasked with preparing the future workforce of professionals, researchers, and technicians along with small and medium business entrepreneurs. These disciplines continue to function in spite of quarantines and obstacles to learning imposed by new learning modalities.

My work as a university professor

The nationwide quarantine was announced on Sunday, March 15, and became effective on Monday, March 16, the same day the first semester of 2020 was to begin. However, that same week, the government decided that the start of classes would be postponed until March 30, as a preventive measure just in case the pandemic worsened. Shortly after, the government decided that classes would start on April 6 via virtual methodology, given the long-lasting quarantine spreading throughout neighboring Spanish-speaking countries in the region.

Professors and teachers had only two weeks to transform courses from traditional to virtual instruction. At first, this task generated a lot of resistance from teachers who said that education would not be the same under the new teaching system. Added to this was the extra work throughout the year, with little time to prepare for them.

Long-term quarantine and fear of contagion

At the same time, long-term quarantine and the fear of contagion caused many people to lose their jobs. For private universities, this implied students would stop studying if they did not have experience with computers and the internet. The university would also suffer significant economic effects, which would reduce the hiring of lecturers given budget cuts. Professors with tenure, as in my case, would also have to handle more students in their courses using the new methods in the virtual classroom.

Teaching plans in the university

I had to plan in two weeks for the opening the virtual courses and learn about Zoom and other discussion platforms (which, luckily, I already knew). Likewise, there was a big problem of adapting methodology because virtual courses are not the same as face-to-face courses when it comes to using materials. University lectures rely on copying the professor's notes and visiting the library, while virtual education requires core and supplementary materials to be made available online or in printed packets for everyone.

I had to change the methodology of talking alone on Zoom, at least for the first half of the meeting. Although I had in mind a communication dialog, in most of my classes the only person I had in front of me was myself. I could not see any reaction as individual student computer cameras were turned off. I realized I could not force them to turn on their cameras as they may be in compromising situations – some might be sharing the room with small siblings, or girls might become victims of harassment by others taking screenshots of them.

I also report anecdotes of students really doing other activities during class time: People joining late, or accidentally turning on cameras while in the bathroom or bedroom. I had the opportunity to speak with students who were frustrated and stressed and who asked to talk to me after class because they felt alone. Virtuality can generate strong professor-student contact but over time, contact among students may falter. Finally, for the last half of class time, I livened up my lectures. As I am a big fan of listening to the radio, I decided to make my classes into an artificial radio or television environment, where I could be more dramatic and entertaining. I was not expecting to get feedback from students, but I learned that they really enjoyed the change. The same was true for other colleagues who changed their teaching delivery style.

When planning the courses, we were also asked to take into account that not all students faced the same resource conditions. Some faced internet access problems because national networks connecting houses were fully saturated or because more family members were using the internet at the same time. There were also cases of students sharing computers or students studying on their mobile phones.

Faced with this situation, we were asked to generate as much material, such as recorded videos or audios, which could be made available offline and viewed anytime. This task created great tension for me – I wanted to make "professional" audios, but during the day there was too much noise in the house because of my playing children. I couldn't record audios very well at night either because I didn't want to wake anyone up in the family. All this generated stress while working from home. As it turned out, most students did not even want these materials, let alone the group chat room discussion forums we recommended to them. They wanted immediacy, preferring Zoom classes that imitated traditional lectures from before.

Likewise, much more work was required for the teachers because we had to create materials – videos, audios, special texts, and grade weekly forums – and provide feedback. We had to prepare all of these, while our children at home required supervision with schoolwork given to them through the new virtual schooling.

Trying to make our classes as professional as possible also meant some economic investment not covered by the university. Professors had to buy better computers and professional microphones, and improve their internet connection, chairs, and tables, to recreate their office at home. We had to cover these costs individually. This situation was repeated in almost all Latin American countries. It was the same experience everywhere I traveled as a visiting professor in Colombia, as most Latin American countries have the same restrictions when using online education.

Life as a family under COVID-19

Something I consider positive about the pandemic was being able to spend more time at home with my children. My children were nine and six years old at the start of the pandemic. I spent more time with them on weekends because during the week I sometimes had to give master classes at night. On certain days, they only saw me early in the morning. Later, I was able to take them to school once face-to-face classes resumed. Spending time with them was very important, but it also meant that the tasks entrusted to the school were now taken over by the parents. Although my eldest son could be more autonomous in class, the youngest who just started his first year needed more help. Thus, I invested in mornings to accompany him in his virtual classes. This added six to seven hours per day dedicated to the primary education of my two children.

Spending more time with my children was exciting at first. Over time, however, it became complicated to balance helping and teaching them subjects such as math or reading and stress management. Fortunately, my children's classes were learning from autonomy, and the teachers also asked us not to intervene too much. We had one function – encouraging motivation – so the children would concentrate on the class. Because of this, I had to continue investing time in my new work as a disciplinarian, where it became difficult for me to read my own texts because of the continuous distractions of my children.

As a result, I invested my time in cleaning around the house, and I developed a new passion for reading manga, or Japanese comic books. After the children's classes and household chores, I would spend an hour doing physical exercises with my sons, since it was believed that COVID-19 generated more risk in overweight people. Minding my children and preparing virtual lectures took much time and affected the quality of my sleep.

Working from home also allowed us to work from other cities on occasion. For Christmas 2021, we visited my parents in Cusco, my hometown. We spent more than a month there because we needed to quarantine for 14 days to avoid contaminating my parents. As Cusco is a tourist place, the pandemic hurt its local economy.

Access to food and other household products

Despite the lockdown, people had the right to go out an hour a day to go shopping for meals. Because we could not use cars, we had to shop in nearby places. Initially, the government did not support or allow delivery services.

People started to hire taxi drivers, which were allowed to work in different markets and supermarkets because supermarkets were not prepared for delivery. Delivery services were subsequently enhanced and the consumption habits of Peruvians radically changed, including eating at home rather than at a restaurant or café. Remote purchases also improved Peruvians' confidence in online shopping and the use of electronic transfers, virtual wallets, and courier services. As online commerce for importing goods from abroad got reinforced, the confidence that Peruvians had to buy locally expanded internationally.

Final words

The COVID-19 pandemic caused many human losses, but it also forced changes that otherwise would not have occurred in Peru:

Peruvians understood the importance of monitoring improvements in public health. Virtual e-economy based on online purchases and payments was strengthened. The fact that Peruvians have been forced to shop remotely generated confidence and use of them. Likewise, the use of electronic wallets has become widespread and is now the second means of payment, even displacing credit cards.

In higher education, a new virtual teaching based on video calls was strengthened. This format, which was initially rejected, is now highly valued by graduate students who study and work because they do not have to waste time in traffic. *Perorado* students, those who like discussion and discourse, are the norm in Peru. As they prefer face-to-face education that goes well with their highly interactive social relationships, it will be interesting to see how well virtual education adjusts to this cultural style.

The COVID-19 disaster of 2020–2022 will mostly accelerate the use of new technologies, especially in higher education. However, this will still take time, given the long tradition of lecture-style formats of the past. With advancing technologies tied to the internet and the many apps, the next generations will improve their ability to study independently and search for information and ideas useful to their work and lifestyles. This is the change that the pandemic may slowly force on the education community, as knowledge increases throughout Peru and the world.

8
THE PLIGHT OF VIRTUAL EDUCATION IN INDIA DUE TO COVID-19

Anwesha Pal

Introduction

The pandemic has affected the world as we know it in multiple ways. When the lockdown was announced toward the end of March 2020, all educational institutions across India were closed abruptly and students were sent home from residential facilities. Not only has the ongoing pandemic changed our social relationships, but it has also brought a giant change in the university pedagogy and ways of learning. Face-to-face classroom teaching has been replaced by virtual modes of teaching. These online classrooms have become the closest equivalent to the physical classrooms now. This chapter discusses the subsequent discovery of shocking realities relating to socioeconomic divides and student responses to new methods of teaching and learning, especially in the higher education sector.

This chapter offers the point of view of a teacher teaching at one such university in India. It shares my experiences and observations relating to the challenges the pandemic has brought about in the higher education sector in the country.

Primary and secondary education

On May 25, 2022, a leading national daily in India claimed, based on a government's survey research report, that a substantial number of students found learning at home over the pandemic to be burdensome. Learning at school scored better and was more accepted compared to online learning from homes. This survey was undertaken by the government when studying schools across countries that dealt with primary and secondary education (Chandra, 2022). The government survey used school students as samples for studying the acceptability of online classes and teaching systems.

DOI: 10.4324/9781032690278-10
This Chapter has been made available Under a CC-BY-NC-ND license.

To the utter dismay of the surveyors, they learned that 78 percent of students found online learning burdensome, whereas 24 percent of students reported they lacked any digital devices at home. One of the factors that contributed to the students' dislike for learning online was that there were no designated hours for learning or doing assignments. There were too many assignments that were doled out on a rolling basis. Moreover, help from peers and one-on-one teaching by the instructors in classrooms were missing, which made it difficult for the students to better understand the technical subjects of mathematics and science as opposed to language-based subjects.

Tumpa's grit

A story closer to home reflects a very different world devoid of all things modern – the life of a domestic helper who was "unfortunately" blessed with two daughters in a world where female children are considered a burden to society.

Tumpa is a domestic helper who works as a cook and cleaner for the households of her rural locality, which is fast transforming into a semi-urban setup. She makes around INR 5,000 (less than USD 50) a month and yearns to give her two daughters, ages 11 and 6, a good education that would enable them to get a job suited to their qualifications. She does not want them to be limited to domestic help when they grow up. Tumpa lives with her in-laws in a slum area on an unclaimed government land near the railway tracks. Her husband used to work as a daily laborer at a local factory office but was indicted for stealing money. Since then, he worked for a local *tuk-tuk* owner and drove the electric *tuk-tuk*, making less than INR 2,000 a month (less than USD 20). (Tuk-tuk is a small motorized transport vehicle.) Since the pandemic struck, the *tuk-tuk* did not fetch the bare minimum either.

Tumpa was not allowed to enter households out of fear of spreading the virus from her rural locality. Because of this, Tumpa lost her job a few weeks into the pandemic. One of her most difficult challenges was how to keep her daughters in school. The government-aided schools where her daughters, Rachna and Ria, go to are situated far away from their home (see Figure 8.1 and Figure 8.2). An even bigger challenge was access an electronic device that would help them learn at home, just like most of her classmates were doing. Because Tumpa's meager remuneration stopped due to the pandemic, her daughters were left without access to education altogether. To add to their woes, Tumpa and her husband are illiterate and cannot help educate their daughters.

Right to education

The Right to Education Act of 2009 in India stresses on free and compulsory education for all between the ages of 6 and 14 years.[1] When enacted, this legislation is perceived to be one of the most progressive pieces of legislation that

80 Anwesha Pal

FIGURE 8.1 Rachna at her school (Ghoshpara Nishchinda Balia Vidyapith, government-sponsored).
Source: © Anwesha Pal

FIGURE 8.2 Ria at her primary school in the village.
Source: © Anwesha Pal

would help prevent school dropouts and alleviate literacy rates and education among citizens of the country. The right to education is recognized as a basic human right. In India, it is read as a facet of Article 21 of the Constitution, which discusses the right to life and is later codified in Article 21-A.[2] However, the very purpose of this Act got defeated when Tumpa's children had zero access to electronic devices or means to continue their education online as government-aided schools switched to new methods of teaching during the pandemic. The government did not come to the rescue by ensuring their children's education. There were also no organizations that could gainfully employ Tumpa or her husband.

While a lot of students across the country were happy attending classes online from the comforts of their home, Tumpa's daughters faced a different reality. Early mornings until late afternoons were filled with household chores and fieldwork, whereas late evenings were about cooking and cleaning the homestead and the vicinity. With absolutely no one to help with their studies, they were staring at a bleak future and a life ahead that could possibly be worse than their mother's.

Role of the state bureaucracy

The government's decision to dole out electronic tabs and smart mobile phones came in much later, around the end of 2021. The unsystematic distribution was politically motivated by impending elections in the States. Rural parts were prioritized over semi-urban or urban localities, and not many got the benefit of the progressive government schemes to ensure learning among students even during pandemic (Edugraph, 2021; Pradesh & Adityanath, 2021).

During the pandemic, students and their guardians were worried about the levy and payment of fees, which were otherwise payable in advance. In some areas, state governments stepped in and tried regulating the fees of private and government schools. These notifications were challenged by the respective schools. The challenge was unsuccessful before the High Court. However, for the *Indian School, Jodhpur and Anr. vs. State of Rajasthan and Other*, the Apex Court ruled on March 8, 2021, partly in favor of the schools and partly in favor of the students.

The Supreme Court considered the aspect of financial slowdown and directed payment of school fees in installments, and further directed no coercive measures on account of nonpayment of fees. Such a direction was a welcome step. Requests for adjustment in fees owing to the pandemic had been argued by premier law school alumni. These policy decisions and the High Court decisions were much awaited. These bold steps by the judiciary helped the masses have faith that the judicial system, as the third organ of the state, would always stand tall.

As the pandemic created a new normal, our lives began slowly adapting to the new change. Its effects can be seen across societies, irrespective of strata, age, and number. With the onset of the pandemic in March 2020, the world seemed to have reached a dead end. Roads and buildings were forsaken, and the lecture halls of universities and schools alike remained empty, waiting for their students to return. For the first few months, life had come to a grinding halt. Over time, new software came to be developed and implemented widely, so as not to hamper prescheduled academic sessions. Universities and schools started training themselves with technological advancements and software so that even during the deadliest of times, a teacher can conduct lectures and educate their students, including those sitting far away in remote localities.

University education

In elite national law schools of the country, such stories are hard to come by. Nonetheless, they do exist. I speak from my experience as a teacher at the WB National University of Juridical Sciences, Kolkata, which is a premier national law school of the country. Because legal education offered at national law schools is expensive, the gentry is generally the class that can afford to spend a comparable sum of money on the education of their children. However, over the years, good Samaritans have been benevolent in giving scholarships to students who harbored the dream of studying at these hallowed halls of legal education. However, the dreams of these scholars came to a screeching halt with the onset of lockdown caused by the rising pandemic.

Remote tribal areas without access to online classes

Students who belonged to extremely remote tribal areas in the states of Odisha and Jharkhand were left with no recourse to access their online classes. The privilege of having a continuous supply of electricity, let alone internet connectivity, was still a distant dream in such villages. As a teacher lecturing at educational institutions situated in urban localities, access to electricity and internet never posed a concern so big until now. The socioeconomic divide that existed before is now glaringly evident from the realities of the different sections of the society.

Pandemic and technology: a quagmire

The winter semester was nearing completion when the 2020 lockdown was declared and evaluation had to be conducted online. Apart from the aforementioned discovery of a student residing in a tribal area of the country, I also witnessed a student residing in the war-stricken area of Kashmir. This student could not appear for his scheduled online *viva voce* exam. When I received a request to reschedule the date of his exam, I learned that he had to travel 60 kilometers, all the way from the district of Sopore, to access a computer center located in a relatively urban district of Srinagar. Kashmir has been mostly under a constant internet blackout with intermittent connectivity for a few days over the past five years or so (Ellis-Petersen, 2020). In such a situation, the student had to make the best out of what was given to him. This was the first time I became aware of such shocking socioeconomic divides as a teacher.

Although subsequent evaluations remained smooth in their execution, online classes were faced with some hiccups. Among the students reeling under the conditions of no internet connectivity, there were those who were based in such remote areas and were visually impaired too. In these trying times, these students found it harder to continue their education.

Based on news reports, with the rising number of students belonging to families with limited means, many sought to end their lives, owing to disillusionment about continuing studies via electronic gadgets and heavy data packs of internet. Such a reality seemed to have been hitherto unknown to many academicians of the country. Adding to the students' woes, unfortunately, conventional scholarship amounts did not seem to cover these expenses either.

We must all remember the age old saying, "Man invented computers/digital gadgets and not the other way round," and the phrase, "Necessity is the mother of all inventions." When the pandemic brought everything, including the education sector, to a grinding halt, the usage of internet and computers, along with different platforms (Zoom, Webex, Microsoft Teams, Google Classrooms), became prevalent. The unthinkable happened: Classrooms reached the doorstep of students and vice versa. However, in developing economies like India where much of the population lives below the poverty line, the luxury of digitization and its effect on the education sector could not be seen or perceived during the pandemic. Urban schools and colleges have equipped themselves with modern infrastructure. However, in rural areas where even the basic necessities of life like electricity is a problem, internet and digital classrooms seemed to be a far-fetched dream for many children.

Digitalization engulfs the country

The traditional method of classroom teaching was the preferred method of instruction among teachers. However, the pandemic changed it all. The new normal became the norm – the digitization wave had engulfed the nation. I was teaching international investment law and company law in the winter semester of 2020 when the pandemic struck and students had to return home. The university and the teachers were faced with a task – how to conduct teaching and examinations within a reasonable time frame.

Classes were now being conducted over online platforms such as Microsoft Teams and Google Meet. The traditional method of taking attendance by roll call stood rejected from the system. Rather, marking of attendance became easy, as participants were recorded by means of an additional extension to the software.

In my experience, the quality of education did take a back seat, as the syllabus and the focus area turned more theoretical than practical when engaging students. This is especially true in legal education where a practical-based approach is required for students' growth, which was diminished to a large extent during the pandemic. Further, real-time explanations with examples and group discussions started fading away. Classroom discussion for interesting topics became mundane once intense dialogue was limited.

Student absenteeism

The biggest bane of the system was absenteeism of students on the online platform. The common problem faced by teachers like myself was the lack of face-to-face interaction with the students. With the videos switched off, it was also impossible to know the actual attendance of students in every class. Class participation declined sharply – only a handful of students asked questions while others chose to stay silent.

The quality of the assessment procedure/examination system suffered during the pandemic. Students were mostly given tasks of writing research projects and theoretical questions, which replaced the system of quizzing them with the practical approach and system. The quality of questions from students was based on random Google searches rather than proper queries based on newspaper reading, blogs, or research articles. Subjects taught at these universities require more practical approach, updated reading, and knowledge based not on just some basic reading but also on news articles, recent amendments, and so on. However, only a handful of students read or asked credible questions. One cannot rule out the problems faced by the visually impaired students. These students were among those who suffered the most in the system.

Solving the quandary

Socrates stated that "The secret of change is to focus all of your energy, not on fighting the old, but on building the new." The pandemic has proved to be a boon for technological advancement in all fields especially in the education sector, as it brought classrooms to the doorsteps of the teachers and students alike. Adaptation of everything new took time and went through teething problems. The entire sector did not suffer due to the physical closure of institutions; rather, online teaching and examinations helped students complete their respective courses and degrees.

The purpose of education has conventionally been the acquisition of knowledge. However, over the years this purpose changed into the ability to acquire jobs. The purpose behind professional education in law is to enable students to apply the laws to various fact situations. Therefore, the entire exercise of classroom teaching predominantly revolves around the discussion of cases and hypothetical fact situations and discussions on the applicability of law. Since law teaching is much more than just imparting theoretical information about law, online classrooms proved to be a disappointment to the extent that they failed to provide the whole experience that students of professional education seek. However, with changing times, instructors need to adapt to the changing needs of their audience.

In the context of legal education, instructors could ask their students to engage in more discussions relating to the application of law to situations they come across in the news. Additionally, the students could send in short opinion writeups after every class to reflect on the learning and takeaways from previous sessions. Although the instructors will have to invest a significant number of additional hours to plan for such classes, they can ensure learning if the students are engaged effectively in the aforementioned ways.

These problems are not restricted to online teaching. The student community in India is by and large grade-centric and not knowledge-centric. It should be a joint effort by the regulatory bodies (University Grants Commission, All India Council for Technical Education, Bar Council of India, and so on), the universities, and students to make education more knowledge-centric than grade-centric. Only then can the problems of absenteeism be taken care of effectively.

Conclusion

In a country like India, online teaching faced more problems caused by COVID-19 than it proved to be a boon. Technically, classes were being conducted and students were attending, but the blackboard teaching, which is the most effective method of instruction, was missing. Instructors were kept busy designing online modules and teaching methods rather than concentrating on students and their approach toward the class. These are my personal experiences relating to several universities in India that are premier in the legal education.

Notes

1 The Right of Education Act provides for the right of children to free and compulsory education until completion of elementary education in a neighborhood school. "Compulsory education" means the obligation of the appropriate government to provide free elementary education and ensure compulsory admission, attendance, and completion of elementary education to every child in the six to 14 age group. "Free" means that no child shall be liable to pay any kind of fee, charge, or expense that might prevent him or her from pursuing and completing elementary education. See https://dsel.education.gov.in/rte for more details.

2 J. P. Unnikrishnan Case, Constitutional challenge querying whether the "right to life" in Article 21 of the Constitution of India guarantees a fundamental right to education to citizens of India; role of economic resources in limiting right to education; interplay between *Directive* Principles and State Policy in the Constitution and Fundamental Rights; and whether the right to education includes adult professional education.

Reference list

Chandra, J. (2022, May 25). 80% students found remote learning burdensome, missed peers: Survey. *The Hindu*. www.thehindu.com/education/80-students-found-remote-learning-burdensome-missed-peers-survey/article65461274.ece

Edugraph. (2021, November 25). *State governments distribute tablets, laptops among students free of cost*. www.telegraphindia.com/edugraph/news/state-governments-distribute-tablets-laptops-among-students-free-of-cost/cid/1840521

Ellis-Petersen, H. (2020, January 5). "Many lives have been lost": Five-month internet blackout plunges Kashmir into crisis. *The Guardian*. www.theguardian.com/world/2020/jan/05/the-personal-and-economic-cost-of-kashmirs-internet-ban

Pradesh, U., & Adityanath, Y. (2021, December 20). UP govt to distribute free smartphones, tablets to 10mn final year students. *Business Standard*. www.business-standard.com/article/elections/up-govt-to-distribute-free-smartphones-tablets-to-10-mn-final-yr-students-121122000543_1.html

9
COVID-19 AND TERTIARY EDUCATION

Experiences in Lesotho institutions

Tsotang Tsietsi

Introduction

The COVID-19 pandemic affected most aspects of human life. It spread across continents and countries at record speed. We were confronted with terrifying global and regional statistics on infection rates and death rates on a daily basis. Coupled with these threats, people also had to contend with rising unemployment and food insecurity. Mental health took a battering. Isolation and fear became prevalent personal and social themes.

Not to take anything away from other spheres of human activity, one sector that was hit hardest by the pandemic was tertiary education. The first response of governments scrambling to contain the spread of the virus was to impose lockdowns. These varied in their degrees of strictness. However, the common goal was to contain human movement, especially the conglomeration of people. The concept of social distancing made its way into common parlance. One of the lockdown measures adopted by governments across the world was the closing of educational institutions. This chapter focuses on tertiary education, in particular. The World Bank defines tertiary education as "all formal post-secondary education, including public and private universities, colleges, technical training institutes, and vocational schools" (The World Bank, 2021, para. 1).

While COVID-19 mitigation measures affected tertiary education across the globe, the impact was hardest felt in developing countries, and, in particular, in the least developed countries (LDCs). Their institutions were far less prepared to move teaching and learning outside of the traditional classroom setting. This was due to various challenges, such as deficiencies in their financial and human resources, and low levels of digital technology acquisition.

DOI: 10.4324/9781032690278-11
This Chapter has been made available Under a CC-BY-NC-ND license.

This chapter discusses an array of experiences during the suspension of teaching and the commencement of remote learning. It also formulates conclusions and advances lessons learned.

Lockdown and the suspension of education in Africa

The United Nations Economic Commission for Africa (UNECA) reports that in 2020, at least 42 African countries imposed lockdowns of various degrees to contain the spread of the COVID-19 virus (UNECA, 2020). Lesotho was no exception. In fact, it was the last African country to notify a case of COVID-19.[1] This was in May 2020. However, in a bid to be proactive, a decision had been made to enforce a lockdown on March 29, 2020. This was executed under a declaration of a state of emergency.[2] Tertiary institutions were closed as a consequence thereof.

At first, the closure was anticipated to be a short-term intervention. The hope was that schools would reopen and there would be an opportunity to make up for the lost teaching and learning time. The lockdown was initially announced as a 21-day intervention. However, this period was extended. And then it was extended again. The collective hope of easily making up for lost time was dashed.

The school closures meant that students studying in Lesotho institutions had to vacate their residences and return to their homes – sometimes in remote areas of the country, or even outside of the country. This was a collectively traumatic time for students and their families, school administrators, and instructors. All constituencies had to navigate extreme uncertainty.

For students, the main uncertainty was whether they would acquire sufficient knowledge to perform well enough to pass and move to the next level, or to graduate and enter the job market, which was similarly groaning under the weight of its own uncertainties. Families were anxious about how to provide support and whether they would have a financial burden of paying for an additional academic year if the year was declared lost. School administrators were pressured to think steps ahead on how to recover every day lost to the heavy lockdown. Instructors were plagued with uncertainty about whether they could complete the syllabus in a genuine and accountable manner.

Commencement of remote teaching and learning

Challenges for institutions

As it became apparent that the lockdowns would continue for an indefinite period, tertiary institutions were forced to consider – many of them from scratch – how to continue operations in this unexpected dispensation. At that point, few institutions in Lesotho had any experience, or even ideas, of how to move teaching and learning online. This was unlike in the developed world, where many institutions had already, for years, been offering remote programs or blended learning and had the financial and technological resources to smoothly transition from face-to-face learning.

Some institutions in Lesotho lacked effective learning management systems (LMSs) by which teaching could commence. Those who managed to acquire such systems were not out of the woods. They had to ensure the skills of learners and teachers to use such systems. Some faced hostility from traditionalists within the institution who refused to acquire these new indispensable skills. Some LMSs had very basic functionality and could not provide essential services, such as platforms for video or audio engagement. They could only be used for file sharing (e.g., study notes, test and assignment scripts, and feedback). This compromised student-lecturer engagement.

As the world extolled the power of platforms such as Zoom and Microsoft Teams for teaching, institutions in Lesotho looked on bleakly. They faced financial challenges of acquiring sufficient licenses of these programs that could effectively allow teaching and learning to continue. The general country infrastructural deficits (breakdowns in electricity supplies and interruptions to ICT) added a further dimension of difficulty. While some institutions in more developed countries were able to provide data for students to access such technologies, those in Lesotho could not afford to. In a country where the majority of the population live in abject poverty and where communication services are expensive, this meant that students could not afford to access such platforms.

Challenges for lecturers

Instructors were forced to navigate unchartered waters in terms of effective lecture delivery. For decades, they had taken it for granted that teaching meant standing in front of students and speaking to them while being able to see their faces and read their body language and facial expressions for clues of comprehension. One could write difficult words on the whiteboard or blackboard for students who were unfamiliar with the terms. One could illustrate and use visuals in support of verbal explanations. A good lecturer knew how to spark interest and get debates going in class. How were these to be emulated in remote learning, especially with the challenge of not being able to use platforms such as the ones mentioned earlier?

The WhatsApp platform

As a compromise, many lecturers resorted to exploring the WhatsApp platform for teaching and learning. WhatsApp was originally designed as a messaging and calling platform. Its makers probably never anticipated that it would, one day, be used to impart very technical discipline-specific content in a higher-education setting. The benefit of WhatsApp was that almost every tertiary student and lecturer already had an account and was proficient in using it. Anyone could create a group and add members and *voila*, a virtual classroom could be simulated. Some classes used video and audio technologies to engage on the platform.

However, digitalization required relatively greater data usage than just chat exchanges. Unfortunately, this resulted in some students not being able to afford the download costs and missing out on content, which was discrimination in access. Confinement to textual exchanges was cheaper, but it led to slow engagement because it depended on varied typing speed. The effectiveness of a textually based lecture also depended on a lecturer's skills in rousing interest, which was not easy. Lecturers would type, give information or pose questions, and be met with eerie virtual silence.

Connected to the above was the breakdown in the student-lecturer relationship. Pre-Covid students and lecturers engaged face to face in lecture halls. Many times, students would accompany the lecturer as they walked back to their office, continuing to engage with them on the lecture content. There were dedicated consultation times when a student could knock on a door and have access to their lecturer. Or a lecturer could call a student to come for further assistance. These opportunities allowed good relationships to develop. This assisted students' learning and lecturers' confidence that their students were learning and coping with their studies. Without these interactions, students didn't know their teachers and vice versa. This compromised the formation of healthy and supportive relationships for the benefit of the students.

Another oft-cited challenge was dealing with safeguarding against academic malpractice. Lecturers tell of the frustration of being confronted with work that was obviously copied from the internet or duplicated among students. There were rumors that senior students, or even graduated students, wrote tests and assignments or even examinations for current students. How was a lecturer to mark work that they suspected was not a true reflection of the student's ability?

Challenges for students

Educational institutions are not for administrators. Nor are they for instructors. They exist for the benefit of students. Even in the best of times, it is difficult to study in an LDC institution. Capacity constraints will always affect learning, whether it is teaching capacity, access to learning materials, or the adequacy of space – classroom space, leisure space, accommodation, etc. The imperative of remote learning introduced as a result of the COVID-19 pandemic made an already tough situation even more dire for students. In discussions with students about their personal circumstances, some shared that they were attempting to learn in homes that were not conducive to studying.

These are small homes, with little to no dedicated working space – homes with limited or no power, let alone internet connectivity. Some found themselves dividing their time between studying and assisting with household chores – responsibilities they never would have had to participate in had they remained at school. Lecturers were trying to teach students who were tired, stressed, demotivated, angry, and full of fears.

One student recounted how physical activity had always been a coping strategy for dealing with stress related to studying. However, with restrictions to the freedom of movement, they were mostly confined indoors, thus affecting their mental fortitude for their studies. Another argued that they couldn't muster the motivation to study as hard as they might have, had they been learning face to face. They couldn't feel the same level of pressure and knew that lecturers became lenient in sympathy of their circumstances.

Before the COVID-19 pandemic, examinations were written physically, in an examination hall, and were completed within a matter of hours. During the pandemic, examinations were moved online. Usually, students were allocated far more time than usually granted in a physical exam (sometimes up to three days), to cater for variations in typing speed, etc. While this ameliorated some of the challenges posed by online examinations, challenges were still experienced. For example, sometimes examination scripts would need to be submitted using the LMS.

At submission time, the system could get overwhelmed by the volume of submissions and, at times, crash. Students would allege that they had submitted, and yet instructors could not find any evidence of such submissions. Time extensions didn't protect students from experiencing other difficulties, such as power cuts or internet disruptions during examination time. One accounting student reported that taking examinations online during the pandemic was expensive, isolating, nerve-wracking, and led to her constantly feeling anxious and depressed.

While remote learning introduced potential difficulties for all students, there are three categories that were hardest hit.

First-year students

The transition from secondary education to tertiary education has always been a challenging one. For many students, this is the first time they will be away from home. They will need to learn financial responsibility and navigate complex adult relationships. Stay-at-home measures heavily affected them. Many missed the important milestones of being freshmen – the orientation, clubs, sports, and other social essentials of being a student. Besides these personal losses from the educational perspective, the learning curve from high school to tertiary education is steep. First-year students need to acquire the discipline to study consistently. They need particular care and attention to support their transition. They need teachers they can see and peers they can relate with in school.

Due to COVID-19, students were not able to interact effectively. They were taught online, and it was hard for them to figure out how to use the school's LMS. The LMS did not work well during examinations, which led to failing marks. The pressure was too much for some students who experienced stress and depression.

In Lesotho, educational institutions witnessed a sharp decline in the performance of those who were in their first year during the pandemic. A current third-year student reported that she was in first year when classes moved online and that, although she passed her first year, she realized in year two that there were a lot of gaps in her knowledge – things that she should have learned in first year. She barely managed to pass second year. She had to take supplement courses in her third year because she lacked fundamental knowledge that she should have acquired in her first year of studies.

Students in practical disciplines requiring laboratories to support learning

In institutions of higher learning, there are students who need to attend practicums. There are programs where observation is a significant part of the learning experience or where students need access to complex and technical equipment to conduct experiments. Teaching such students in the absence of these environments, apparatus, and techniques robbed them of experience and skills foundational to their further education. A BSc agriculture student regretted missing out on practicum activities in his second year of studies. He said that he wished he had participated in a mushroom cultivation project, where he would have learned how to produce mushrooms for sale and health benefits of the crop.

Postgraduate research students

Postgraduate students in the developing world rely heavily on physical libraries. The libraries in Lesotho are not typically well-stocked, but they are often the only available option for those learning in institutions who cannot afford costly subscriptions to a range of online databases. Therefore, the pandemic meant that research students were deprived of their primary source of material, as well as a venue for focused and dedicated study. This lack of access delayed their progress with their research.

Research studies are also already incredibly isolating ventures. Unlike with taught programs where students study in common with their peers, have common assessments, and can collaborate with research, such as in doctoral studies, a research student pursues their own independent study. Being on this individual path in the midst of a global pandemic is even more isolating than it would be under normal circumstances. This is partly why, during COVID-19, completion rates and dropout rates rose for doctoral students. So, too, did incidents of seeking mental health support. One student who was in the final year of her doctoral studies recounted how her anxiety and depression became unsustainable, and she sought medical intervention (medication and therapy) to be able to cope.

With this mix of challenges, some institutions in Lesotho experienced student protests. Some of the issues that students were complaining about were the instability of learning management systems, institutions' insistence on trying to adhere to the ordinary calendar during the unprecedented times, and a general lack of support for their studies.

Despite the above, some students point to positive results of learning during the pandemic. For example, one student named Mafusi shared how her lecturers would upload voice notes on lecture content. Students could download and listen to them at their convenience, and as many times as they needed to. She said she preferred this over a traditional lecture where a lecturer might say something once and move on. If a student missed it, that was it.

Conclusion

While all countries were challenged by the impacts of the COVID-19 pandemic, no doubt developing countries, and LDC as a subset, suffered the most. This was due to their limited resilience in coping with external shocks. The education sector was one of the hardest hits. While some may argue, "What is the value in speaking about the pandemic so many years down the line?" the truth is that its effects still linger and will continue to be felt for some time. It is important to recall what all stakeholders (administration, instructors, and students) had to cope with and to heed important lessons going forward.

Lessons learned

There was tardiness in reacting to the imperatives of the pandemic. This was followed by imperfections in executing mitigating arrangements to continue teaching and learning. This led to all stakeholders experiencing unsavory consequences. However, tertiary institutions fought hard to maintain their mandate of delivering high-quality education. Many lessons have been learned.

Cooperation: Students, guardians, lecturers, government, and university administration need to work together to devise interventions.

Student-centeredness: There is a need to recognize that not all students experience challenges uniformly. Each needs different interventions that best address their peculiar circumstances.

Flexibility: This is needed as opposed to the rigidity of trying to stick to the status quo, regardless of all evidence that it is no longer tenable. An example is the use of WhatsApp for teaching.

Creativity: Financial endowment will always widen the ambit of possible supportive measures. However, institutions adopted other nonfinancial measures of support, such as leniency with timelines, etc.

Innovation: As a result of challenges with academic malpractice, many institutions have started using anti-plagiarism software.

Adaptability: By learning how to teach outside the traditional forms, some lecturers and students prefer to continue with remote learning, despite restrictions being removed.

Notes

1 However, it is not certain whether this was because it was indeed the first case or if it was due to deficiencies with testing capacity.
2 See Chapter 16, "Change and continuity: COVID-19 and the Philippine legal system" by Antonio G. M. La Viña and Chapter 18, "The goldfish and the net(work)" by Nicole Mazurek.

Reference list

UNECA. (2020, May 9). *COVID-19 for Africa: Lockdown exit strategies*. www.uneca.org/covid-19-africa-lockdown-exit-strategies

The World Bank. (2021, October 22). *Higher education*. www.worldbank.org/en/topic/tertiaryeducation

10
LIFELONG LEARNING AS A POWERFUL FORCE IN THE POST-PANDEMIC WORLD

William Loxley

Introduction

Striving for social mobility is a powerful human activity. As a billion young people seek their place in the world, education is a major route to connect talent to skill to career to livelihood to self-awareness. Education enhances upward mobility based on greater choice from more opportunities. Because modern societies need people who can communicate and solve problems, traditional education was born. Over time, this process allowed the most privileged to take advantage of education relative to the disadvantaged. Education now has become the major avenue connecting mobility to social position for highly motivated individuals. This process is a valuable factor in determining future career goals for marriage, career, and family. Upon retirement, senior citizens also require lifelong education to continue mental growth and emotional well-being.

As a senior citizen and former teacher, my observations of the pandemic on youth and seniors have been relentless. Education has taught me humans are pretty much the same everywhere. They want what is best for themselves and their family. Unfortunately, most people fail to achieve these goals. My work has taken me around the globe working in education. In the Maldives, I helped the government set up the national university that provided locally delivered degrees. In Pakistan, I assisted the government in establishing schools for primary girls. In Sri Lanka, I helped the authorities provide computers in secondary schools. I analyzed school dropout rates in Egypt and South Asia; diversified education in Colombia and Tanzania; and provided technical expertise in government research institutes in Indonesia (Loxley, 2016). Each of these interventions helped individuals aspire for a better position in life: for girls to receive basic skills, secondary school students to learn skills in scientific thinking, and higher education to offer merit-based paths to leadership.

This chapter highlights education's role in reshaping each new generation; predicts changes in future education models that address fairness in opportunity; and links individual progress to the "metaverse of everything" (Stinchcombe, 1968). Teaching and learning are central to basic and secondary education, and new technologies online and offline are critical to the new education model. In higher education, academic study focuses on fields of study that lead to opportunities in pursuing careers. Likewise, in well-financed and staffed school systems, most children stand a good chance to develop these skills. Yet in poor countries, this prospect is a pipe dream for many. As the digital world alters the learning process, student-centered focus will use adaptive learning oriented to individual need. The pandemic showed how poorly we equip our future generations to live in the digital age, especially in poor nations where the process is more severe (Reimers, 2022).

Consequences of the 2020–2022 pandemic

The pandemic disrupted the timing and sequence of the lecture and discussion method, as well as the process of learning principles to theories and understanding not only the "what and how" but also the "why" of problem-solving. For example, according to the staff of Little Clarion Montessori School in Manila, the pandemic affected primary school children's problem-solving and social skills once they returned to school (LCMS, n.d.). Teachers, too, were left to fend for themselves in the digital world of teaching, where most of them were unprepared for. The pandemic showed the flaws of the old education model as it translated to socioeconomic mobility in the worlds of work and job markets, especially as they affected select groups of students.

The recently retired and elderly also suffered disruptions as they sought lifelong learning to better understand their personal well-being and awareness of life's meaning. All age-related groups were severely disrupted during the pandemic as schools, offices, and public places were closed. Students had to learn virtually on the internet; work from home forced workers to zoom and connect digitally; and the elderly were left isolated and highly susceptible to COVID-19, as more than half of all Covid-related deaths were among those older than 65 years.

However, in spite of problems, the pandemic taught one thing to the world. All institutions and personal development will involve ICT and require a thorough upgrade of digital know-how into the public sphere in education, public health policy, and business. Service provision for equitable opportunity to learn and grow wiser will be determined greatly by an individual's ability to pay. Unfortunately, the developing world of half the world's population will have to stay content in employing the old education model until financing and skill shortages allow the new model to take hold. The pandemic has altered the future by accelerating the push for the metaverse that is now taking shape around the world.

The coming post-pandemic world

The world population is approaching eight billion, half of which live in urban areas and loosely represent citizens of the modern economy. Among this modern urban sector of four billion inhabitants, 500 million are ages 1–5. From this, one can extrapolate the population of ages 6–22 to be 700 million, ages 23–65 to be 2.3 billion, and ages 65 and older to be 500 million. These age cohorts represent life experiences based on learning and knowledge. The remaining four billion inhabitants residing outside the modern sector will continue to struggle to take advantage of opportunities to succeed.

Young people spend over 15 years learning knowledge and socialization skills that will allow them to follow paths for future development. Some paths are academic and vocational while others include adventure, military, drugs, or crime. Another stage of learning involves midlife career, which sees building partnerships, family, networking, career, and creative life based on opportunities availed by the youth. A further group in their mid-60s to mid-80s and beyond look for second-chance beginnings, discovery, travel, volunteering, self-awareness, and hobbies: all promoted through lifelong learning and technical prowess. Education plays a big role to prepare the youth for work and the seniors for lifelong learning.

Reproducing societal knowledge among the youth

On a large scale, schools introduce the young to the world of knowledge and teach socialization skills and etiquette during face-to-face learning. Knowledge prepares students for life experiences by offering choices and opportunities to advance. Learning moves from facts to the higher order manipulation of understanding, applying, and thinking about ideas needed to emote values and behaviors. These skills offer studies in academic disciplines that deepen awareness over time, so that when students reach the early 20s, they will be ready to contribute to society. Once in place, individuals build on this foundation to increase their choices throughout their lives, including achieving financial literacy.

It is fairly obvious to teachers that elementary school children require face-to-face classroom experience to learn and acquire socialization. At the same time, teaching should move from teacher to student-directed learning, as in Montessori-style hands-on experiences with computers and mobile science tables, and lots of learning materials. Students can then gather a wide array of facts and create concept understandings and applications. By secondary school (ages 13–17), students are inclined to have more group study, classroom interaction, and debate, including gaming exercises.

Finally, by university level (ages 18–21), students aim for multiple approaches employing lecture, group projects, individual programmed study, use of hardware, and technical assistance in problem-solving exercises. Each of these approaches requires adaptive learning with quick turnaround and problem solving using real-world examples. Then too, higher education models require greater use of out-of-class practical application, engineered by highly integrated IT platforms connecting all facets of the learning process, from teaching and team collaboration to discussion and evaluation.

If education is designed to provide upward mobility to each new age cohort, then the pandemic really hurt the chances of the youth to purse their careers (OECD, 2021, 2022). Not only were students barred from schools, they lost out to small business recruitment for over a year or two. Now it seems the organization of business is being realigned – although demographics favor the young as a result of the "great rotation" and job resignation and changing jobs among older workers during the COVID-19 lockdown.

During the pandemic, those younger than 17 years were restricted from leaving home. Staying at home forced them to employ the internet to communicate with friends and acquaintances. As a result, those without computers and access to an internet connection suffered unduly and fell behind cognitively and emotionally in their studies and social maturity. School administrators and teaching staff were forced to redesign their teaching protocols to accommodate new learning alternatives. Because of the pandemic, new approaches are now beginning to take shape.

Strengthening knowledge among the elderly

As a *bona fide* senior, I know that the newly retired, especially those beyond the age of 65, were devastated as health monitoring ceased for those living alone or residing in retirement communities where visitors were not allowed. The isolation and lockdown made the loss of mobility more noticeable, and the fear of contagion and death more real. Like Holocaust survivor Viktor Frankl, crises require a deep search for life meaning and purpose. COVID-19 was not kind to the elderly, who were the most susceptible given their lower immunity levels to the virus and immobility due to poverty (Debanjan, 2020).

As the pandemic slowly subsided in late 2022, the elderly needed to regain confidence, reset their schedules, eat, sleep, and socialize more often in public. Seniors must physically reinvigorate personal exercise to maintain the ability to walk and keep an active mindset to strengthen seeing and hearing about new ideas. Life is like riding a bicycle for seniors. To keep a steady balance, they have to keep moving and have direction. To do this, they must learn to volunteer, travel, and actively support programs that help the most in need.

Many of these activities can be carried out under the auspices of the community and require time and energy. Staying healthy is a priority challenge to remain active and contribute to society. Social policy requires lifelong learning through active participation in social programs highlighted by reaction to the pandemic.

Society also needs to reorganize health services for the senior community and monitor prevention, as the percentage of seniors increase in the world of tomorrow. Seniors today must lead the effort to educate their ranks and carry out lifelong learning among each other. In the 21st century, people 75 years old can look forward to an average continued life of another ten years (more for women, less for the poor). Seniors need to take advantage of their time left to improve themselves and help others. They should not entertain a purposelessness that prevents human development, like waiting for Godot.[1]

Education enters the metaverse

A primer is needed to show how transformation will work in the fields of education and work that blend technology to learning in the metaverse (UNESCO, 2022). The metaverse is defined as a virtual-reality space that uses the internet to interact with a computer-generated world. It should benefit education and training, career planning, and mental and emotional well-being. Education will gain from new modes of delivery, psychology of learning, mentoring, group dynamics, and machine learning all tied to augmented reality (UNICEF, 2018). Career guidance will also help individuals manage choices through simulations and mentoring in the digital world of multimedia. Self-awareness and social etiquette will evolve from virtual simulation while engaging with artificial intelligence that allows the young to gain experience and the elderly to know themselves better.

The internet has recently been converted to the worldwide web, which in turn is now creating the metaverse that spans the real and virtual worlds (Loxley, 2022). The transition is technology driven through platforms, hardware, and software that generate self-awareness and understanding via advanced learning technologies. The world of the avatar will be replaced by the three-dimensional hologram that looks and talks like oneself. There are incredible possibilities for technology working in the metaverse to help individuals and groups meet their potential to gain upward mobility and a sense of purpose.

The youth will greatly benefit from the metaverse, based on the use of simulation explanations carried out in remote learning and long-distance personal relationships using augmented reality applied to the real world. Augmented reality enhances life in the real world to better control outcomes. Face-to-face and machine learning may come together in elementary and secondary schools to teach basic information from daily experience in spite of the complex link between personality and learning style.

Perhaps socialization skills that create human awareness and reinforce self-confidence can be taught in virtual reality better than in real life. This happens as augmented reality allows the learner to adjust situations by repeating them over and over. As machine learning takes shape, it may provide many virtual alternatives to choose from and then assess individual prospects for success based on personal characteristics. The metaverse may provide the youth with awareness of the many connections between skill development and career options for future lifestyles. These connections will allow them to improve and perfect their strengths. As multiverse research opens many new avenues for study and research, the education profession needs to prepare for the coming change.

For the elderly, the metaverse allows them to revive memories of times past and relive them the same or differently. It also allows them to look forward in time to imagine anything they desire. Through volunteering, outward-facing externally directed activities become a worthy goal to project self-awareness and skillsets onto others. Caring for the environment, concern for public safety, and helping the youth are things that can leave a permanent imprint on the future of each older generation.

Lifelong learning among the elderly projects their knowledge and awareness onto life's meaning. It is directed inward, guiding each person on a journey of understanding their life that serves the common effort. The pandemic pointed out inequalities in societal institutions and reminded us that everyone deserves a fair chance to contribute to society. Changes to education in the digital world are sure to improve the power of opportunity.

Note

1 "Waiting for Godot" is an expression that describes a situation where a person is waiting for something that will probably never happen. This is based on Samuel Beckett's play, *Waiting for Godot*, where two men wait in vain for a third one to arrive.

Reference list

Debanjan, B. (2020). *The impact of Covid-19 pandemic on elderly mental health*. John Wiley Public Health Emergency Collection.

Little Clarion Montessori School. (n.d.). *Welcome page*. https://littleclarion.com

Loxley, W. (2016). Assessing the impact of knowledge on development. In V. Jakupec & M. Kelly (Eds.), *Assessing the impact of foreign aid*. Elsevier.

Loxley, W. (2022). *Knowledge networks in the exponential age*. www.linkedin.com/pulse/knowledge-networks-exponential-age-william-loxley/?trackingId=6Gzr4OYFS5mKQyE20f8Kig==

OECD. (2021, June 15). *COVID-19 pandemic highlights urgent need to scale up investment in lifelong learning for all, says OECD*. www.oecd.org/newsroom/covid-19-pandemic-highlights-urgent-need-to-scale-up-investment-in-lifelong-learning-for-all-says-oecd.htm

OECD. (2022). *Delivering for youth*. www.oecd.org/coronavirus/policy-responses/delivering-for-youth-how-governments-can-put-young-people-at-the-centre-of-the-recovery-92c9d060/
Reimers, M. F. (Ed.). (2022). *Learning from a pandemic. The impact of COVID-19 on education around the world*. Springer. https://link.springer.com/content/pdf/10.1007/978-3-030-81500-4_1.pdf
Stinchcombe, A. (1968). *Constructing social theories*. University of Chicago Press.
UNESCO. (2022). *Guidelines for ICT in education policies and masterplans*. https://unesdoc.unesco.org/ark:/48223/pf0000380926
UNICEF. (2018). *ICT for learning process & tools, Vol. 1 – schools*. www.unicef.org/esa/media/7536/file/ICT-Learning-Process-Tools-Vol-I-Schools.pdf

PART III
Public policy and risk management

11
INTRODUCTION TO THE POLITICAL SECTOR

Suresh Nanwani and William Loxley

Introduction

Risk management in public goods and policymaking is designed to resolve social problems effectively (Granovetter, 1985; Turner, 2022). COVID-19 showed the public sector's inability to address black swan events in highly volatile circumstances. Governments' failure to use science and commonsense values to educate and communicate them worsened the effect of the pandemic.

The COVID-19 experience teaches us to maintain strong institutions (Ye et al., 2020). It appears past relevant policymaking experience goes a long way to get governments through pandemics. Good organization assures trust in the common good based on transparent and accurate data, education of public needs, and international, bilateral, and national institutions that stand ready to provide public services. Organizations can then promote the common good over individual rights in times of crisis through equitable access to resources, effective economic productivity, and revitalized public health and education systems (Hazelkorn & Gibson, 2019).

Command and control skills useful in master planning and good governance based on data and research is the domain of government, along with NGO assistance. The public sector's role is to ensure rule-based choices over anarchy, balancing group considerations over individual rights. Governments create global institutions such as WHO, IMF, WTO, and ILO to manage health capital flows, trade, employment, and immigration. Based on data analysis, public institutions set immediate protocols and schedules needed to address disasters, inform and educate the public to gain legitimacy and trust in the fight to overcome adversity, and mobilize people and resources. These tasks are done best with open communication, diversity of thought in ideas

DOI: 10.4324/9781032690278-14
This Chapter has been made available Under a CC-BY-NC-ND license.

and actions based on past experience, and science aimed at a well-educated citizenry (Book Authority, 2023). Every nation needs to provide these multilateral institutions undivided attention (Bismark et al., 2022).

Public sector policy requires highly educated and trained experts to corroborate and forecast social, economic, political, and cultural trends (Adler, 2022). Data-driven policy implies employing quantitative reasoning interpreted alongside a sense of emotional IQ, to address public needs and show that actions taken are for the common good. Big data permits digital monitoring to help explain trend lines. When pandemics arise, big data can quickly address problem solving while ICT platforms can be tweaked to monitor and communicate information. This is best seen in managing education, health, social services, R&D and business activities, and support for the arts that produce empathy and humanity in populations (Ryan, 2023).

Because of the COVID-19 catastrophe, public policy has to realign public health systems. This is because institutions nearly collapsed under the weight of hospitalization numbers due to a lack of resources such as oxygen tanks and beds. Reorganizing public services requires an educational rethink (Logiudice et al., 2020). In addition, vaccine research has to be supported and financed by governments globally to hasten production of reliable new drugs and therapies. Many governments were unable to coordinate the health systems because of under-supported staff expertise or funding and lack of training. Then, too, government policy adjustments were required to aid those most affected by the pandemic – frontliners, the poor, and the medically compromised. Management systems and staff training to meet crises needs are essential if society is to respond quickly and effectively to future catastrophes (Davis, 2018).

As more and better knowledge expands throughout societies, digital and lifelong learning about health, retirement, and aging will help better prepare individuals and institutions to become aware of demographic-led trends in society. The youth needs to appreciate adults; adults need to appreciate seniors; and seniors need to appreciate the youth. Personal responsibility and community cooperation are essential in fighting pandemics (Dipankar, 2021). All these learning experiences lessen personal fear of the future and encourage personal wellness, enabling more policy options to fight disasters and preserve human development (Logiudice et al., 2020).

Seven essays

The second theme that the project identified as a key area that COVID-19 impacted around the globe is public policy and risk management during the pandemic. The institutions involved are mainly government and some NGOs in the political and social sectors. Each institution aims to provide services and promote science-based information to support decisions strengthening the common good that protects society (Chan et al., 2014). Issues within this theme include efficient implementation of policy, group coordination, cooperation and trust, law and order, public cost containment, and protection of the public good including public health systems (Duca & Meny-Gilbert, 2023). Chapter topics analyze issues including information dissemination, vaccine and supplies, trust in data, support for international trade and international organizations, public service quality, and support for science research.

Reference list

Adler, M. (2022). *A research agenda for social welfare law, policy and practice.* Elgar Research Agendas.
Bismark, M., Willis, K., Lewis, S., & Smallwood, N. (2022). *Experiences of health workers in the COVID-19 pandemic: In their own words.* Routledge.
Book Authority. (2023). *20 best social policy books of all time.* https://bookauthority.org/books/best-social-policy-books
Chan, R., Lih-Rong, L., & Zinn, J. (2014). *Social issues and policies in Asia: Family, ageing, and work.* Cambridge Scholars Publishing.
Davis, J. (2018). Management education for sustainable development. *Stanford Social Innovation Review, 18.* https://doi.org/10.48558/VZVM-Q853
Dipankar, S. (2021). *Pandemic, governance and communication: The curious case of COVID-19.* Routledge.
Duca, F., & Meny-Gilbert, S. (2023). *State – society relations around the world through the lens of the COVID-19 pandemic: Rapid test.* Routledge.
Granovetter, M. (1985). Economic action in social structures: The problem of embeddedness. *American Journal of Sociology, 91*(3), 481–514.
Hazelkorn, E., & Gibson, A. (2019). Public goods and public policy: What is public good, and who and what decides? *Higher Education, 78,* 257–271. https://link.springer.com/article/10.1007/s10734-018-0341-3
Logiudice, S., Liwvhaber, A., & Schoder, H. (2020). Overcoming the COVID-19 crisis and planning for the future. *Journal of Nuclear Medicine, 8,* 1096–1101. www.ncbi.nlm.nih.gov/pmc/articles/PMC7413231/
Ryan, J. M. (Ed.). (2023). *COVID-19. Individual rights and community responsibilities.* Routledge.
Turner, B. S. (2022). Towards a sociology of catastrophe: The case of COVID-19. In J. M. Ryan (Ed.), *COVID-19: Surviving a pandemic.* Routledge.
Ye, Q., Zhou, J., & Wu, H. (2020). Using information technology to manage the COVID-19 pandemic: Development of a technical framework based on practical experience in China. *JMIR Medical Information, 8*(6), e19515. https://pubmed.ncbi.nlm.nih.gov/32479411/

12
NEW ZEALAND

Global connectivity and digital diplomacy

Tracey Epps

Introduction

New Zealand has been widely praised for its COVID-19 response, especially in the early stages of the pandemic when the country avoided extended lockdowns. In this chapter about the sociology of development, I explain measures taken by the government and reflect on the impact for a geographically isolated nation for which so much depends on its interactions with the rest of the world.

New Zealand's COVID-19 response

Until the inevitable arrival of Omicron, New Zealand was one of a very few countries worldwide that maintained a COVID-19 elimination strategy. During the early stages of the pandemic, the country's geographic isolation helped it maintain cases at a low level. By mid-March 2020, it only had six active cases. Nevertheless, on March 19, 2020, the risk presented by the pandemic led the government, for the first time in history, to close the country's borders to all but New Zealand citizens and permanent residents. Two days later, the government announced a four-level Covid alert system, placing the country under different rules depending on the level of Covid risk.

Initially, the alert was set at Level 2 (Reduce). Domestic travel was permitted, but measures were put in place to ensure social distancing and reduce the virus's spread, for example, a 100-person limit on gatherings such as weddings and funerals. Hospitality businesses were limited to a maximum of 100 people. Shortly after, on March 23, 2020, the alert was raised to Level 3 (Restrict), which set significant limitations on interregional travel and restricted attendance at events like weddings and funerals to ten people (Unite Against

DOI: 10.4324/9781032690278-15

This Chapter has been made available Under a CC-BY-NC-ND license.

COVID-19, n.d.). Hospitality venues had to close (although restaurants were permitted to serve takeout in a contactless manner).

March 25, 2020 – state of national emergency declared

On March 25, 2020, a state of national emergency was declared, and the country moved into Level 4 (Lockdown). This was one of the world's most stringent nationwide lockdowns, with people required to stay at home and no travel permitted, except for necessities (supermarkets and pharmacies remained open) or to undertake safe outdoor recreational activities. Schools were closed, and only businesses considered "essential" were permitted to operate.

The country remained in Level 4 until April 27, 2020. On June 8, 2020, the Ministry of Health announced that there were no longer any active cases of COVID-19 in the country, and the setting went back to Level 1 (Prepare). All restrictions on work, school, sports, domestic travel, and gathering sizes were lifted. However, border controls continued. Immigration ground almost to a halt. Only New Zealand citizens and residents were allowed to enter the country, and they were required to undergo a mandatory 14-day managed isolation and quarantine (MIQ) process. Entry depended on obtaining a booking in a government-approved hotel.

2020 to 2022

For the next two years, New Zealand cycled through various changes to the alert settings. The Four Level Alert system was eventually replaced by the so-called Traffic Light System (still in force at the time of writing). In August 2021, a Delta outbreak in Auckland, the largest city and most important commercial center, resulted in a short lockdown for the whole country, and a lockdown of over 100 days for Auckland itself. This saw Auckland cut off from the rest of the country, with travel in and out of the city extremely restricted.

Early on, there was broad support for the government's response – as reflected in an unprecedented election victory in 2020 for the incumbent government led by Jacinda Ardern. That support dissipated over time, as many wanted to see New Zealand move forward and reconnect with the world. The border closures became increasingly controversial as migrant families were separated (caught in situations where one member of the family had been outside the country when the borders closed), and many citizens and residents living overseas were unable to return home due to lack of availability of spaces in MIQ hotels. (Later, in 2022, the New Zealand High Court would rule in favor of a legal challenge by an advocacy group known as "Grounded Kiwis," who sought judicial review on the grounds that the government had acted "unlawfully and unreasonably" in the design and operation of the MIQ system.)

There was also controversy over the vaccination rollout, which was criticized by many as being unnecessarily slow, with bookings only commencing in July 2021. Eventually, however, the country achieved one of the world's highest vaccination rates, with over 95 percent of the eligible population receiving two doses. Vaccination mandates were announced in October 2021, with workers in public-facing non-essential services (e.g., hospitality, gyms, hair salons) required to be fully vaccinated, and members of the public required to present vaccine passes to access these services.

Despite New Zealand's initial success in keeping Covid levels low, the pandemic eventually caught up with it. By early March 2022, cases in the community had reached a seven-day rolling average of over 20,000. At this time, the government started easing restrictions, announcing that the mandated use of vaccine passes would no longer be required at most businesses and venues, and that border restrictions would be eased and the MIQ system phased out. From April 13, fully vaccinated Australian citizens were allowed to travel to New Zealand with no isolation period, and as of August 2022, the border was fully reopened.

The impact of the COVID-19 response

As a small country geographically distant from much of the world's population, New Zealand is strongly dependent on international connections. There are between 600,000 and one million New Zealanders living overseas (Stats NZ, 2020b), while over one quarter of residents were born overseas (Stats NZ, 2020a). People have long relied on the freedom to travel in and out of the country by air and have enjoyed extensive flight options. As noted earlier, the personal impact of the border restrictions was correspondingly significant and often tragic.

International connections are also critical to New Zealand's economy and security. New Zealand is an export-dependent economy, with over 20 percent of its gross domestic product (GDP) made up of exports, and one in four New Zealanders' jobs dependent on exports (NZFAT, n.d.b). New Zealanders need to work hard to stay visible and relevant to the rest of the world, whether it be business people traveling to attend trade fairs and visit customers, or diplomats seeking to maintain influence in international processes.

It was inevitable that COVID-19 would have a significant impact on business, not only because of the supply chain issues that have bedeviled businesses everywhere but also due to the strict border restrictions that effectively prevented business travel. At the beginning of 2020, New Zealand exporters were not alone. The world was shut down, and nobody was traveling. There were some advantages for New Zealanders in this situation – when everyone is on

Zoom, geography ceases to matter. It's when some can travel and others can't that the real difficulties begin to be noticed and stay relevant.

International business

From the beginning of the pandemic, the New Zealand government played a key role in working to keep international business flowing. Early in 2020, it was involved in several initiatives aimed at keeping supply chains open, including a joint declaration with Singapore, Australia, Brunei, Chile, China, Laos, Myanmar, Nauru, the UAE, and Uruguay, affirming commitment to ensuring supply chain connectivity. It also joined 42 other World Trade Organization (WTO) members to affirm support for the multilateral trading system and WTO's role in responding to COVID-19.

Other initiatives included an international air freight scheme to ensure the movement of critical supplies to and from New Zealand. Domestically, the government did the same as many others around the world, providing financial support to businesses, including a wage subsidy scheme. Many of these actions have already been, or will be, phased out as the pandemic recedes.

According to Catherine Beard, executive director of Export NZ (the key advocacy and support body for New Zealand exporters), the closing of borders was a "combination of a blessing and a curse as time went on. It had the advantage of keeping COVID-19 out and allowed the economy to carry on while the country was virus-free, and businesses adapted to using technologies like Zoom to stay connected internationally" (personal communication, July 2, 2022). Many businesses were able to quickly pivot to accelerate digital strategies, often far quicker than they would ever have thought possible before the pandemic (albeit also with increased support from the government).

Overall, New Zealand's goods exporters performed well despite the pandemic. Although demand decreased as the world economy slowed, this proved to be transitory and relatively shallow. Goods exports dropped by 1.7 percent from NZD 60 billion in the year to March 2020, to NZD 59 billion in the year to March 2021. They then rebounded strongly to NZD 65 billion in the year to March 2022 (+10%), supported by high commodity prices for the major exports of dairy and meat products. Yet there were still significant challenges and frustration with the strict measures. As Beard explains,

> As time went on, the Government made some mistakes (such as being slow to get vaccines and RAT tests) and the elimination strategy resulted in regions being locked down, which caused huge stress to businesses. The only thing that stopped complete catastrophe was the Reserve Bank flooding the economy with money which the Government handed out as wage subsidies to keep people employed and businesses on life support. The closed border was also a huge frustration to exporters and to our ability to

get key technical skills into New Zealand that were needed to keep critical businesses operating. And as the rest of the world started opening up, our exporters were missing the people-to-people meetings that are required to win new business.

(personal communication, July 2, 2022)

The most significant economic impact was on the services sector. Prior to COVID-19, tourism and international education were among New Zealand's most important export earners. The sustained border restrictions hit these sectors hard, with revenues falling over 50 percent in the year from March 2021.

Digital diplomacy

Businesses were not the only ones who had to accelerate their digital strategies. Officials also had to learn the art of "digital diplomacy." As a small country, New Zealand is highly dependent on a functioning multilateral rules-based system. This system is still undergoing multiple challenges in present times, of which the pandemic is only one. But for officials, COVID-19 presented a unique challenge – how could they continue to engage and make a difference from thousands of miles away, and in a time zone incompatible with much of the world?

Asia Pacific Economic Cooperation meetings

This challenge was met head-on with some notable success. Nowhere was this more evident than in New Zealand's hosting of the 2021 Asia-Pacific Economic Cooperation (APEC) meetings. New Zealand successfully hosted APEC's first fully virtual year. Between February and November 2021, there were over 350 formal APEC virtual meetings, along with more than 100 testing sessions, informal meetings, and bilateral meetings on the periphery.

While there were challenges, officials also reported new opportunities that emerged from the digital format. For example, because delegates only had to log in rather than travel internationally to attend a meeting physically, New Zealand was "able to get the right people in the room. We could also hold a wider range of meetings, such as the Informal Leaders' Retreat in July" (NZFAT, n.d.a, para. 27). The success was also evident in the outcomes of the meetings themselves, including the Aotearoa Plan of Action (a guide to adapting to new challenges over the next 20 years), and the conclusion of the Indigenous Peoples Economic and Trade Cooperation Agreement – an agreement developed in 2021 by New Zealand and other APEC economies.

Overall, the APEC year showed how much can be achieved without physical travel. As Rachel Taulelei, New Zealand's Representative on the APEC Business Advisory Council, explained, "being able to work digitally has let us go perhaps a little bit harder and faster than we might otherwise. When you rely on physical meetings, you're hamstrung by time and place, whereas when you're working in a digital environment, there are few to no barriers to your ability to interact and engage" (Sachdeva, 2021, para. 38).

Free trade agreements

Other examples of successful digital engagement also took place in the international trade policy space. New Zealand and the United Kingdom successfully negotiated a Free Trade Agreement (FTA) almost entirely on a virtual basis (although chief negotiators were able to meet in person toward the end of 2021). Much of the negotiation for the New Zealand European Union Free Trade Agreement was also conducted virtually before being concluded in person in June 2022.

An FTA sets the terms on which countries will treat each other when it comes to doing business together – whether importing, exporting or investing – and negotiations typically involve lengthy and technical discussions between large teams of government officials. To have worked through these negotiations without the benefit of sitting down together in the same room, and in time zones 12 hours apart, was a remarkable achievement.

Reflections

As everywhere, the lessons learned from the pandemic are many and varied. New Zealand is a country that cannot easily survive without strong global connectivity, and this reality was highlighted by the pandemic. The last two and a half years have shown that both the private and public sectors can adapt quickly when faced with a crisis, and some incredible outcomes were achieved. Who would have imagined, ten years ago, that an FTA could be negotiated almost entirely through virtual meetings?

Despite the successes, the pandemic has served to highlight the ongoing importance of in-person connections. As Simon Draper of the Asia New Zealand Foundation says, "Many of New Zealand's most successful professional, trade and diplomatic relationships with Asia are grounded in real-life, face-to-face experiences that have taken place in the past few decades" (2021, para. 5). The speed with which business people have taken up the opportunity to visit customers overseas once border restrictions were lifted illustrates the value that many place on those face-to-face meetings.

Time may also reveal lost opportunities as diplomats have remained onshore and missed out on those "meetings in the margins" and conversations they may otherwise have had once the world started opening up. Meanwhile, the speed with which many young New Zealanders are leaving the country to pursue experiences and careers abroad shows that the desire for travel has not dissipated.

A fortress country mentality

The pandemic also brought to bear an undercurrent of a "fortress New Zealand" mentality, with many people worried about the border being reopened, even at a time when other countries had moved ahead to do so. It is to be hoped that, over time, all New Zealanders will embrace an open outlook and that divisions sowed by the pandemic will diminish between those who wished to open up to the world and those who did not. Our connectivity may look different in the future as people take advantage of digital technologies and travel more judiciously (with carbon emissions in mind).

Government and business will have to grapple with difficult questions as they seek to balance the need for personal interactions with climate change impetus and budgetary restraints. These questions will create ongoing conversation and debate, but there is hope that a balance can be found that will allow New Zealand's continued open and fruitful engagement with the world.

Reference list

Draper, S. (2021, August 30). *Connecting more important than ever in the Covid era*. Stuff. www.stuff.co.nz/business/opinion-analysis/126214127/connecting-more-important-than-ever-in-the-covid-era

New Zealand Ministry of Foreign Affairs and Trade. (n.d.a). *Delivering a year of virtual meetings*. www.mfat.govt.nz/en/trade/our-work-with-apec/delivering-virtual-meetings/

New Zealand Ministry of Foreign Affairs and Trade. (n.d.b). *Trade recovery strategy 2.0*. www.mfat.govt.nz/en/trade/trade-recovery-strategy/trade-recovery-strategy-2-0/

Sachdeva, S. (2021, November 9). APEC's digital diplomacy in a time of Covid. *Newsroom*. www.newsroom.co.nz/apecs-digital-diplomacy-in-a-time-of-covid

Stats NZ. (2020a). *2018 Census data allows users to dive deep into New Zealand's diversity*. www.stats.govt.nz/news/2018-census-data-allows-users-to-dive-deep-into-new-zealands-diversity

Stats NZ. (2020b). *About 100,000 New Zealand residents travelling overseas*. www.stats.govt.nz/news/about-100000-new-zealand-residents-travelling-overseas

Unite Against COVID-19. (n.d.). *History of the COVID-19 alert system*. https://covid19.govt.nz/about-our-covid-19-response/history-of-the-covid-19-alert-system/

13
BRAZIL IN CRISIS MODE

Institutions in times of uncertainty

José Guilherme Moreno Caiado

The Pandemic and the world

Setting the context for the pandemic

A few decades ago, the construction of a multilateral (and peaceful) world seemed to be just a matter of time. The 1990s and early 2000s were full of references about the success of multilateralism and the rise and spread of democracy. Typical examples include the end of the Cold War, the accelerated EU expansion eastward, the Chinese and Russian membership to WTO, and the entry in force of countless bilateral and multilateral trade and investment agreements.

Despite much optimism, however, the benefits and costs of globalization were not equally distributed among countries nor among the domestic groups within the countries. Whether coming from a nostalgic left inclination or an unashamed right propension, dissatisfied domestic groups grew and so did their political representation in key economies, such as the United States and the United Kingdom. As a result, the years that preceded the pandemic were to become famous for the rise – to the highest levels of policymaking – of anti-global, conservative, and religious thinking.

Uncertainty and the reaction of states

In a way, therefore, a certain "trend to unilateralism already existed pre-pandemic," especially in areas of international law such as international trade and security, as well as related international institutions. While critics to multilateralism were already trending by 2020, the pandemic nevertheless "took governments by surprise" and escalated the problem, with countries taking a series of uncoordinated responses to mitigate the effects of COVID-19. As mentioned by the interviewees, the pandemic very quickly "posed a major challenge to each and every state." Within a few weeks after the first reports of the virus, governments, suddenly facing the risk of an outbreak within their borders, were forced to "mobilize resources and ensure access to key inputs and materials" in every possible way.

The fast and decisive reaction of governments was probably a necessary element to slow the contagion chain down and protect human health. However, the changing environment of the pandemic and the lack of consolidated scientific knowledge engendered a fertile soil for uncoordinated trade measures designed to ensure access to such key inputs. On the one hand, one could observe a surge in trade liberalization measures. These measures were designed to facilitate the import of relevant medical products, such as masks and ventilators. Interestingly, due to the scarcity of key inputs and the global reach of the pandemic, governments of producing countries, wanting to ensure supply of their domestic markets, perceived this wave of liberalism as an opportunistic attempt that threatened to divert input to the export market, to the detriment of the domestic market.

As reaction, a counterwave of restrictive trade measures, such as export controls, were used by governments to secure key inputs. Such export controls were "justified on a declared state of national emergency, especially in the beginning of the pandemic." These measures added up to the pre-Covid wave of protectionism. As mentioned by one interviewee, "The imposition of unilateral measures like tariffs, sanctions, and other additional duties [in the United States] had increased since Trump's Administration in 2018, and most of these measures remained in place during the pandemic and to date in Biden's administration." All in all, it seems that in the presence of such an "atypical situation," the "capacity for coordination between states decreased," adding extra pressure to an already wounded multilateralism.

The role of multilateral institutions

In these moments of high uncertainty, international institutions should play an important role in fostering collaboration. However, the setup and dynamics of different international organizations differ significantly, and so did their responses to the coronavirus crisis.

The interviews suggest that at least three reasons play a role in explaining the differences in timing and substance of the response of different organizations. First, some organizations had already been weakened in the years before the pandemic. For instance, WTO's decision-making system can be partially held responsible for the organization's incapacity to conclude significant multilateral deals and resolve organizational issues like the Appellate Body vacancies.

Second, the organizational setup of these institutions also became an issue, with one interviewee expressing the view that "not all of these organizations were designed to operate in a crisis mode" and their decision-making models ended up influencing their deliverables. For example, "the World Bank mobilized . . . resources very quickly and deploy[ed] them so that countries could respond to the pandemic."

The International Monetary Fund (IMF), an institution that deals with governments in moments of crises, quickly established credit lines and put in place other measures to secure governmental funding for COVID-19-related actions. According to IMF, governments from 90 countries had made use of one or more IMF COVID-19-related programs, for a total amount of over USD 170 billion by the end of 2021. Consensus-oriented organizations with a larger membership, such as WTO, "needed more time and . . . only managed to be on [a] pandemic responsiveness package [two years later] in 2022." This is due to the fact that consensus is arguably a lengthier decision-making process in most cases, especially when a group as diverse as the WTO membership needs to be onboard with a decision.

Finally, the political disputes of the pandemic spilled over institutions such as WHO, an organization that was "very pressured by some actors who tried to influence the narrative of the pandemic." WHO responses to the crisis, however, were praised by all interviewees, especially regarding its "leadership in the COVAX initiative, which is a vaccine sharing mechanism that has several contributing countries." In a way, therefore, the pandemic "helped increase awareness of what [the organizations] do and highlight their weaknesses" so that these organizations "have come out of the pandemic with a renewed sense of purpose."

The interviewees share the impression that WHO managed to resist political pressures to make one or other country responsible for the pandemic. Instead, it managed to focus its efforts on multilateral initiatives to mitigate the effects of COVID-19, due, in great part, to the technical and scientific focus of its staff and network of supporting domestic institutions and individuals. This conclusion makes sense if one considers the linkage between politics and growing unscientific belief in the times immediately before the emergence of COVID-19.

Interestingly, more technical and therefore less political organizations can be criticized for their lack of representativeness (after all, technicians are not democratically accountable in the same way as politicians are). Nevertheless, it will be interesting to see in further research, in the next years, whether institutional setups used at WHO may have contributed to this success in comparison to other organizations.

Similarly, the World Bank, an institution with a strong mandate that could quickly respond to the challenges, may be compared to a more "member-driven" organization that depends on consensus for decision-making. An example of this is WTO, whose responses needed a longer period of negotiations to be put in place.

COVID-19 and Brazil

Brazil in context

The COVID-19 responses by international organizations were different in substance and timing due to, among others, the institutional setup of each organization and its capacity to foster cooperation in times of a global health crisis. At the end of the day, cooperation and collaboration between states are "dependent on sovereign countries," their characteristics, their immediate needs, domestic institutions, and long-term interests. In order to understand Brazil's reaction to the COVID-19 crisis, it is relevant to understand the country's defining characteristics and current political and economic context.

Brazil is one of the largest and most populated countries in the world. It has a land mass almost as large as the United States and a population of over 200 million people. The country has a great deal of natural resources, a powerful agricultural sector, and is the South American regional industrial and service power. Yet the saying goes that Brazil is the country of the future and will always be. This almost sarcastic prediction echoes the country's cycles of prosperity followed by economic and social crises, never fully realizing the potential of its natural, technical, and human resources.

In the last of these cycles, the Brazilian GDP rose to the point that the country became the world's sixth economy. At that time, things seemed to be working out for Brazil and its socially oriented government. Public and private investments were abundant, and poverty was consistently being reduced by a combination of economic growth and social programs. For international companies and investors, Brazil was the place to be, with the country hosting the Football World Cup in 2014 and the Summer Olympics in 2016. However, Brazil could not sustain the high level of economic growth. By the end of 2014, the country entered an economic recession that would drive millions of protesters to the streets and culminate, a few years later, in the impeachment of government and the beginning of political turmoil. Shaken by a series of corruption scandals, Brazil would have witnessed by 2020 the temporary imprisonment of two former presidents and dozens of senior members of parliament and local governments.

This political chaos cleared the path for the election of a rather unknown and virulent right-wing congressman and marked the beginning of a politically belligerent government. It was under this fragile political and economic background that the COVID-19 pandemic hit the country. The lockdowns, the closure of borders, and the disruption of supply chains posed further stress tests to the already unstable economy, leading to hundreds of thousands unemployed and a severe rise in poverty levels.

Structural disorientation

Domestic politics was also disoriented. The country experienced four ministers of health within a short period of time, and local governments often had to operate in the absence of a clear federal strategy. The federal government often downplayed internationally recognized prevention practices such as mask mandates, social distancing, and testing, thus failing to set an appropriate tone from the top and raising the levels of uncertainty among the population. More than once during the pandemic, the health system collapsed, with shortages ranging from oxygen to ICU beds. Two years later, Brazil lost over 680 thousand people directly or indirectly due to the COVID-19 virus, a bitter second place only behind the United States. Yet, despite these "several economic, social, and political consequences," the interviewees expressed a somewhat respectful view about Brazil's conduct of the pandemic. Why?

Brazilian institutions

The answer seems to be in the Brazilian institutions rather than in the Brazilian government. While many politicians downplayed the isolation and mask mandates, technical-oriented institutions, at either the federal or local government level, offered a safe space for qualified scientific debate that would influence policymakers during the pandemic. Moreover, these institutions preserved communication with international organizations such as WHO. As a result, the policies they proposed were better aligned with international standards on how to fight COVID-19. These institutions provided for "domestic surveillance system and . . . were able to mobilize quickly to scale up vaccination for a lot of people." Despite the constant credibility attacks against vaccination coming from the President's Office, the institutions managed to continue operating with a high level of technical expertise.

For example, the Oswaldo Cruz Foundation (FIOCRUZ), an institution associated with the Brazilian Ministry of Health, signed an understanding with a British company in 2020 so that it could produce vaccines in Brazil. Similarly, Instituto Butantan, an institution affiliated with the São Paulo State Secretariat of Health, not only engaged in international collaborations to locally produce vaccines developed outside Brazil, but also undertook efforts to develop their own COVID-19 vaccine. These initiatives show that certain institutions recognized the risks associated with COVID-19 and made relevant efforts to put remediation measures in place, despite the often critical view of the Brazilian federal government.

The position of the Brazilian government, though not capable of suppressing all Brazilian initiates against COVID-19 in cooperation with the international community, seems to have at least weakened the image of Brazil abroad, in a way that "Brazil could have been a much better role model during the pandemic." For one interviewee, "the pandemic happened in a time that Brazil was not as active in the international arena as it used to be, and this was very apparent in global discussions about common responses." Why was Brazil not as active as it used to be?

Brazilian diplomacy is recognized by its "coherent position of defending the international order and defending multilateralism" irrespective of the current administration. When coming to power, Bolsonaro nominated a highly controversial diplomat as minister. This career diplomat often spoke against a perceived left-oriented globalism and seemed to be in a position to undermine the traditions of Brazilian diplomacy. In the end, the institution seemed to have prevailed. The ministry was removed, and Brazil recalibrated its diplomatic tone.

In a way, therefore, "the major trend in Brazilian foreign policy that has been preserved throughout the various administrations and in this sense the pandemic did not represent a watershed, but only a crisis that was faced with successes and mistakes with all countries." However, this does not mean that the pandemic acted "as substantively changing Brazil's position on the international stage."

Conclusion

This chapter discussed the Brazilian response to the COVID-19 crisis against the background of tensions between multilateralism and unilateralism in a global setting. Three Brazilian experts were interviewed and shared their opinion on topics related to Brazilian politics, trade protectionism, and the role of politicians in shaping policies and institutions, especially in moments of crisis. It can be concluded that the response of international institutions to the COVID-19 crisis was influenced by their institutional framework, especially when confronted with the unusual level of political pressure of the early years of COVID-19, and the trade and political tensions already in place before the pandemic.

The analysis shared by the interviewers points out to the idea that certain institutional settings were better designed to operate in moments of crisis, for instance, by quickly implementing mitigation measures (IMF and World Bank) and defining a clear strategy resilient to political pressures (WHO).

This framework is also applied to understand the responses of Brazil during the crisis, in terms of both international cooperation and domestic strategy. Although Brazil has historically been a supporter of international cooperation, it was not immune to the political and economic tensions in place before the pandemic. Already living in a scenario of economic crisis during the second decade of this century, the largest country in South America entered into a political crisis shortly before the COVID-19 emergency. Under economic and political instability, the Brazilian government would hesitate, to say the least, to devise and implement an effective strategy to protect its population.

While hundreds of thousands were dying in connection to the virus, the Brazilian government insisted in measures that exposed, rather than protect, the health of the population, at the same time promoted ineffective drugs and sabotaged the vaccination campaign. Despite all that, certain Brazilian institutions with a strong technical orientation managed to create scientific knowledge and engage in international cooperation for the development of measures to fight the spread of the virus. Institutions that are somehow isolated from political pressure are often criticized for being isolated from democratic aspirations that only politics, through democratic elections, may offer.

However, this crisis showed that politics and scientifically based policies do not always walk hand in hand. It also showed that a certain level of institutional protection was a positive aspect that allowed many institutions and their employees to continue their work of fighting the COVID-19 crisis.

14
LEGAL PRACTICE IN KENYA

Embracing automation and e-judiciary

Leyla Ahmed

Introduction

The nature of court proceedings has been characterized by in-person participation. It takes one to watch just one episode of *Suits*, a US legal drama set in a fictional New York law firm, to understand why we lawyers love what we do, more so in court. Whereas most movies highlight lawyers as theatrical, the truth is we are storytellers, and we like to tell our clients' stories effectively. Perhaps this explains why the traditional legal practice has been slow in the uptake of technology. However, an accelerated uptake became imminent in the last couple of years.

The COVID-19 pandemic has impacted many sectors, the legal profession being no exception. The most advanced and necessary change that came with the pandemic is the introduction of virtual courts. As human beings, we like to live in our comfort zone. Because change is scary, the new norm was not readily accepted in some parts of Africa. In Nigeria, for example, there were two cases that were filed, challenging the constitutionality of virtual courts on the grounds that under sections 36(1), (3), and (4) of the Nigerian Constitution, court proceedings should be held in public. Although the court had ruled that the two suits were premature, it held that virtual courts are not unconstitutional (Emudainohwo, 2021).

The confirmation of the first COVID-19 case in Kenya led to the scaling down of most government services, including service delivery in the judiciary. The focus of this chapter is the introduction of e-judiciary and how it changed the practice within the legal profession.

The practice pre-Covid: a day in court

The legal practice has been characterized by paper-based procedures as well as physical attendance in court. My typical day would start with coming to the office before 8:00 a.m. As our offices are not far from the High Court, I would start my trip to the court at around 8:30 a.m., armed with the files I needed for the day and accompanied by the pupil – a student undertaking their pupillage program, which is the final stage of training before one can apply as an advocate in Kenya. Ideally, the courts are supposed to start at 9:00 a.m. However, at times there are delays. Woe unto you if the judge's list has a long list of matters, which is the case most of the time, and your matter is among the last ones. You would have to sit through until the end. Even when you had just a mention, you could be in court until past midday, at times even standing, since the number of seats available inside the courtrooms could not accommodate the number of court users.

There are days I would have two to three matters in different courts at the Milimani Law Courts – a petition in the constitutional division on the third floor, another one in the commercial and tax division on the second floor, and another one in the Employment and Labor Relations Court, which is in a separate court building. Running in between different courts was the order of the day. The number of times I have addressed court almost out of breath is countless. I must say, running back and forth between courts was a great exercise for me! It was always a pleasure seeing learned colleagues chatting along the corridors of justice as they waited for their matters to be called out, exchanging pleasantries and discussing interesting cases they were handling.

The other advantage of the physical court was the fact that you could always request a fellow lawyer to assist you with your matter as you handle other matters, in what we refer to as "holding brief." We had a restaurant directly opposite the court. It was the norm to see lawyers heading out to grab a cup of tea or coffee after their court session in the restaurant, as they discussed issues pertaining to their practice or random topics, ranging from politics to life in general.

Documents were also filed physically. First, the lawyer had to submit his documents for assessment on how much court fees are payable at the court registry. Second, once the amount was assessed, he would go to the bank to make payments, where he had to deal with the long queues. Once the court fees were paid, he would then go back to the registry with the receipt so that they could stamp the documents. Only then would the documents be considered as filed. Once the documents were filed, the lawyer would proceed to the offices of the opposing party for physical service of the documents. Some of the offices where lawyers had to serve the documents were outside Nairobi, meaning we had to send a court process server to effect the service. The entire process was time consuming. As a result of the physical attendance at the court registries, the registries were congested with people and files.

The practice during Covid

The measures of court excellence, which are adapted from the International Framework for Court Excellence (IFCE), provide the hallmark of an effective court and include access to justice, expeditious disposal of cases, court files integrity, and case clearance rate. In the following paragraphs, I shall discuss how COVID-19 impacted the court system and what interventions were put in place to ensure the effectiveness of the courts was realized.

Electronic case management system

The judiciary under the Judiciary Transformation Network (2012–2016) aimed to create an e-judiciary framework that would make ICT an enabler of its transformation programs. It aimed at automating judicial operations through:

a) establishing an electronic case management system;
b) establishing an SMS inquiry system to inform members of the public about the status of their cases;
c) digitizing court records;
d) installing teleconferencing facilities;
e) mainstreaming the use of electronic billboards in the courts; and
f) ensuring the digital recording of proceedings and transcription.

The first COVID-19 case was reported in Kenya on Thursday, March 12, 2020, by the Ministry of Health (MoH). As the country was placed under lockdown, there was cessation of movement between different cities. On Sunday, March 15, 2020, the National Council on Administration of Justice (NCAJ),[1] whose composition includes the Chief Justice (CJ), Cabinet Secretary responsible for matters relating to the Judiciary, the Attorney General (AG), the Director of Public Prosecution (DPP), and the President of the Law Society of Kenya, met to deliberate on the implication of the announcement by MoH. A press release was issued that contained an administrative and contingency management plan to mitigate COVID-19 in the justice sector.

To ensure that there was continuity in the dispensation of justice while controlling the spread of the pandemic, NCAJ instituted several measures. As NCAJ is a policy body, its responsibility is to ensure that the functions of all the sector players are coordinated during this period. All non-essential staff were directed to work from home or to take a leave. All staff older than 58 years were also directed to work from home. Proceedings in the courts were suspended for two weeks with immediate effect. During this period, the courts were to continue handling matters brought under certificates of urgency and taking plea for serious cases.

On March 20, 2020, Chief Justice David K. Maraga (2020) issued practice directions for the protection of judges, judicial offices, judiciary staff, other court users, and the general public from the risk associated with the pandemic.

As directed by the notice, the courts may make use of teleconferencing, videoconferencing, and other appropriate technologies to dispose of any matter. Parties were directed to serve court documents and processes through electronic mail services and mobile-enabled messaging applications, as provided for under the Civil Procedure Rules. Service through WhatsApp was permissible. However, save for plea taking and urgent criminal and other matters, nothing happened in the courts for over a month, as operations had been scaled down across the entire justice sector.

In March 2020, the Practice Directions on Electronic Case Management were released through Gazette Notice No. 2357. It required all cases commenced after the enforcement of the practice directions to be electronically filed. However, one could apply to the court to convert their case into an electronic case. The objectives of the practice directions were to provide for the electronic filing and service of court documents, electronic case tracking system, electronic payment and receipting, electronic signature and stamping, and the use of technology in case registration and digital recording of proceedings. The court could exempt people in exceptional circumstances and for a good cause from the requirement of filing their documentation electronically, such as lack of regular access to the internet. The system was officially launched by Chief Justice Maraga in July 2020.

The interface of the electronic case management system

The electronic case management system has two interfaces. The first is a user interface that allows lawyers, law firms, and individuals to register through the e-filing platform. Once registered, documents can be uploaded. The system then generates a digital invoice. Upon payment of the court fees, a receipt is generated. Members of the public can also conduct searches. Digital signatures became acceptable through the Business Laws (Amendment Act, 2020). The second interface is the court interface that allows judicial officers to have access to the court documents.

Merits of the system

Although there were challenges at the beginning, the electronic case management system has improved over the past two years. One can now file documents and serve electronically in less than ten minutes. The electronic payment of court fees has increased accountability and transparency within the judiciary, as it captured the perennial issue of corruption at the court registries. The system sends automated updates through the registered mobile number, enabling one to obtain prompt updates of their cases. Statistics provided by the Deputy Registrar in charge of automation show that as of June 30, 2021, a total of 67,299 cases had been filed through the system, 3,097,000 documents filed in existing cases, USD 7,787,085.42 collected through the e-payment platform, and 8,314 accounts activated in the system.

Demerits of the system

When there is a system downtime, one is unable to file documents. At times the downtime has gone beyond a day, affecting efficiency. Over the past two years, the issue of system downtime has been rampant. However, the judiciary has always responded swiftly by issuing directions for urgent matters to be filed through the various court emails, which are also provided on the cause list – a list of cases scheduled for a particular day in the various courts within the country and provided on the Kenya law website.

Virtual court proceedings

In partnership with the Information and Communication Technology Authority (ICT Authority), the judiciary acquired Microsoft Teams, a video conferencing service, to facilitate upscaling of court operations. This was used for virtual sessions by the Supreme Court, the High Court, and the Magistrates' Courts. The virtual sessions in the Court of Appeal were done through the GoToMeeting platform. The links to the virtual courts are provided on the cause list found on the Kenya law website (www.kenyalaw.org). For matters that are of private nature, the links are shared directly with the parties involved in the matter through email.

Article 50 of the Kenyan Constitution on fair hearing provides that every person has the right to have any dispute that can be resolved by the application of the law decided in a fair and public hearing before the court. Do virtual courts breach the requirement of holding hearings in the public? I think not. The Constitution does not define what a public hearing is. The crucial thing is that what makes a hearing public is the accessibility of the members of the public to the court proceedings (Emudainohwo, 2021). Can members of the public access the court proceedings virtually? Absolutely. As the links are publicly available, anyone can join the court using the link. While the virtual court sessions are open, the public can follow the proceedings, whether they are a party to the matter or not.

Constitutionality of the deployment of technology in the judicial system

In November 2020, Kituo Cha Sheria, a human rights and non-governmental organization (NGO), filed a constitutional petition seeking orders for the reopening of the courts for physical hearing. It argued that the government had failed to consider the plight and realities of many Kenyans, particularly the indigent, self-representing litigants, the marginalized, those unfamiliar with court procedures, illiterate and semi-illiterate, rural, and some urban communities (Makau, 2021). It also argued that the resolution to deploy technology in the judicial system was in contravention of the Constitution for want of public participation and stakeholder consultations. However, the court was of the view that there were circumstances where public participation cannot always be undertaken. Clearly, the pandemic was unpredictable, and the safety of the members of the public, judiciary staff, and prevention of the spread of the pandemic was a priority for the government.

On the issue of whether the introduction of virtual courts was discriminatory against marginalized communities, the court held that it was important to consider the intention of the judiciary in putting in place the new electronic measures – to prevent the spread of the virus among court users and to ensure there is continuity in the dispensation of justice. Whereas the court acknowledged that the new measures may not be easily available to all Kenyans, it found that there was no discrimination against any Kenyan in putting in place the new electronic measures.

Hitches of virtual proceedings

It was a relief when virtual court proceedings were introduced, as one could now conduct their court sessions from the comfort of their homes and offices. However, it was not easy in the beginning, as lawyers settled down to the use of technology. I was before court one day, opposing the grant of an interim injunction, and it was only after I had submitted for a couple of minutes did I hear the judge say, "Counsel, were you muted? We did not hear you." There were times my screen froze while I was in court, and other times when my connection dropped as power disappeared. I had to frantically call my opponent to request the judge to place the file aside to enable me log back in.

The blunders of virtual courts seem to have provided some comic relief to the public. More often than not, lawyers forgot not only to log in with their video off but also to log in with their microphones off. We have had multiple private conversations with people who have logged in and who forgot to mute. The Law Society of Kenya has a dress code that it expects its members to adhere to. At the earlier stages of the virtual court proceedings, we had lawyers in informal attire appearing in court. The judges and magistrates had to remind the litigants that they were required to dress formally even in virtual proceedings.

The strengths of virtual courts

Despite these hitches, the virtual courts have their own advantages. Potential loss of court documents has been reduced as court records have now been digitized. Backlog has been reduced as the court processes are automated. (A case is defined as backlog if it is not finalized or concluded within one year from the date of filing.[2]) The Judiciary's Performance Management and Measurement Understanding Report 2019/2020 indicates that the unresolved number of cases in all courts was reduced from 469,359 in 2018/2019 to 289,728 in 2019/2021.

Between January 2020 and September 2021, a total of 316 Tax Appeals with an aggregate claim of USD 926,346,753 were filed in the Commercial and Tax Division of the High Court in Nairobi. These claims have been dealt with virtually, where 57 Income Tax Appeals have been decided, unlocking a sum of USD 92,731,979 released back to the economy.

It is also time-saving, as time previously used for physical filing and physical service has been drastically reduced. There is also the convenience of attending court virtually. I am able to do other things as I wait for my turn to address the court. Overall, the flexibility and cost effectiveness that come with virtual courts have made virtual court attendance a preferable mode over the traditional method of physical court attendance.

The Commercial and Tax Division, High Court Bar Bench Committee – composed of practitioners before the commercial division, as well as judges and the registrars of the division of which I am a member – had resolved to continue with virtual courts within the division even post-Covid, particularly for mentions for directions, case management, and delivery of rulings and judgments.

Kizuri hakikosi ila

There is a Swahili saying, "*kizuri hakikosi ila,*" which loosely translates to mean that there is nothing so beautiful that does not have blemishes. Whereas virtual courts have been lauded for their efficiency and time and cost saving, there are also some disadvantages. The assessment of nonverbal cues of the witness is limited in virtual courts, making it difficult to assess the credibility of the witnesses. Concerns have also been raised on witness coaching and leading. It also becomes difficult to conduct a hearing in a matter that involves volumes of documentation.

To address some of these challenges, practice directions were developed that required the witnesses to put their videos on when testifying. In the event that the witness is in the counsel's offices, the counsel and the witness should be in separate rooms to avoid witness coaching.

Not only is there a challenge of electricity connectivity as not all areas in Kenya have power, there is also a challenge of internet connectivity. Even a litigant like myself, who is based in the city, has faced an issue of power blackout. How then does the judiciary manage the issue of bandwidth in urban and rural centers? In these areas, the judiciary ensures high availability and reliability of bandwidth by installing both primary and secondary links, which are on fiber with sufficient speed. Whereas the primary link acts as the main link, the secondary link acts as a backup. This guarantees that courts can continue with their operations when the primary link drops. Court stations in rural areas are connected via fiber links, while courts in areas with little challenges of network and internet connectivity are connected using satellite technology.

Alternative Dispute Resolution

Article 159(2)(c) of the Kenyan Constitution mandates the judiciary to promote alternative forms of dispute resolution (ADR) in exercising judicial authority. ADR mechanisms include reconciliation, mediation, arbitration, and traditional dispute resolution mechanisms. The Court Annexed Mediation (CAM), a mediation process conducted under the umbrella of the court, was launched in 2016. The State of the Judiciary and the Administration of Justice Annual Report (SOJAR Report) for 2020/2021 indicates that a total of 767 matters were settled successfully through CAM and that USD 3,160,943.31 was released back to the economy from the settled matters. MAC accredited 126 new mediators, bringing the total number of accredited mediators to 829 at the end of June 2021.[3]

The value of the cases referred to court annexed mediation during 2020/2021 was about USD 58,750,516. The value of the matters settled during 2020/2021 was reduced from USD 37,236,243 to USD 3,160,943. The reduction is attributed to the challenge of holding mediation sessions during the pandemic.

During the 2020/2021 period, the number of CAM matters was reduced as a result of the mitigation measures put in place by the government, including the closure of open courts. Inadequate use of virtual platforms in mediation affected dispute resolution.

Arbitral proceedings

The pandemic also had an impact on arbitral proceedings. In 2020, one of my colleagues lamented how the respondent in his claim had refused to proceed with the arbitration virtually, indicating that they should wait until the pandemic was over. The problem was no one knew how long the pandemic would last. There were a few arbitrators who had already adopted some sort of virtual proceedings even before the pandemic, making it easier for them to adopt the new normal. The pandemic accelerated the adoption of technology in arbitral proceedings as well as virtual proceedings. Arbitral institutions developed protocols and guidelines to be used during virtual proceedings.

In April 2020, the Africa Arbitration Academy launched its Protocol on Virtual Hearings in Africa.[4] The objective of the protocol was to provide guidelines and best practices for arbitrations within Africa, where a physical hearing was impracticable due to health, safety, cost, or other considerations. It was also to promote the use of reliable technology in arbitral proceedings. In October 2020, the Chartered Institute of Arbitrators, Kenya Branch, issued new rules that replaced the institution's 2012 Arbitration Rules. The 2020 Arbitral Rules contained a provision for virtual proceedings, where the arbitral tribunal, on its own motion or on application of either party, could direct the proceedings to be conducted virtually or in a hybrid hearing setting when it was not possible to conduct an in-person proceeding.

What next?

The judiciary is in the process of establishing Judiciary Desks at Huduma Centers with an aim of bringing justice closer to the people. Huduma Centers are physical one-stop shop service centers that provide public services from a single location.[5] This is in a bid to increase access to justice, more so for litigants with no legal representation. They are aimed to offer registry services, customer care, information, and virtual court sessions.

Conclusion

Justice Kashim Zannah (n.d.), Chief Judge of Borno State in Nigeria, noted in his lecture on "Advancement in technology: Signpost or requiem to legal practice," that as technology increasingly makes the world a truly global village, tolerance for sloppy service will be minimal from a populace that is easily aware of better service obtaining elsewhere. In this lecture, he stated that "the legal marketplace of tomorrow cannot be immune from the technology that will permeate the socio-economic fabric of society. Paper-based practices and practitioners will surely be as extinct as dinosaurs are today" (p. 15).

A new era in justice delivery

The pandemic has birthed a new era in the delivery of justice. However, having experienced the virtual court systems, I do not think we are there yet – a place where we can affirm that paper-based practices have become extinct. Take, for example, the recently concluded petitions challenging the outcome of the presidential results in Kenya's general election on August 9, 2022. One of the petitioners submitted his documents in a truck! Voluminous documentation such as those ones would affect anyone's eyesight if they were all to be read from the screen. In addition, internet access and affordability are still major challenges not only in Kenya but in Africa as a whole, especially in rural areas.

According to the CIPESA (2021) brief on "Towards an accessible and affordable internet in Africa key challenges ahead," an estimated 45 percent of

Africans live further than ten kilometers from network infrastructure essential for online education, finance, and healthcare services. Access to reliable and affordable internet is crucial. It is also important that the judiciary looks at how they are accommodating people with disabilities, including the deaf and the blind, in these virtual proceedings.

Equal access to justice should be maintained, no matter the platform used. For example, it is crucial to consider braille for the visually impaired. Until all these factors are put into consideration and solutions are found for these challenges, we cannot boast about a fully virtual court system. It is clear that virtual courts are likely to become the norm and hybrid court system is definitely the new normal.

Notes

1 NCAJ was established under section 34 of the Judicial Service Act, Act No. 1 of 2011.
2 See the Judiciary's Performance Management and Measurement Understanding Report 2020/2021.
3 See the State of the Judiciary and the Administration of Justice Annual Report (SOJAR Report) for 2020/2021.
4 Visit www.africaarbitrationacademy.org/protocol-virtual-hearings/
5 Visit www.hudumakenya.go.ke/

Reference list

CIPESA. (2021). *Towards an accessible and affordable internet in Africa: Key challenges ahead.* https://cipesa.org/?wpfb_dl=482

Emudainohwo, E. (2021). Appraising the constitutionality of virtual court hearings in the national industrial court of Nigeria. *Nnamdi Azikiwe University Journal of Internal Law and Jurisprudence, 12*(1). www.ajol.info/index.php/naujilj/article/view/206730

Makau, J. A. (2021). *Legal advice centre t/a Kituo Cha Sheria & 2 others v. chief justice & 2 others; law society of Kenya (interested party) eKLR (petition E393 of 2020).* Kenya Law. http://kenyalaw.org/caselaw/cases/view/220505/

Maraga, D. K. (2020). *Practice directions for the protection of judges, judicial officers, judiciary staff, other court users and the general public from the risks associated with the global Corona virus pandemic* (Gazette Notice No. 3137). Kenya Law. http://kenyalaw.org/kl/index.php?id=10310

Zannah, K. (n.d.). *Advancement in technology: Signpost or requiem to legal practice* [Speech transcript]. Nigerian Weekly Law Reports. https://nwlronline.com/uploads/Hon_Justice_Kashim_Zannah.pdf

15
THE PANDEMIC AND POST-PANDEMIC AFTERSHOCKS

Whither legal education?

Shouvik Kumar Guha

Introduction: a storytelling approach

The COVID-19 pandemic has begun to be like a thing of the past now, with the new normal gradually returning to the old. Yet, even if the worst of the phenomenon and its global fallout lie behind us, it is unlikely that the pandemic will fade anytime soon without leaving some lasting changes in its wake. This impact has changed our lives, especially in the field of education.

Considerable literature portraying such changes across multiple sectors already exists. Instead of merely reproducing variation of the same, I will adopt a different approach in this essay – involving a sharing of diverse perspectives via stories. Stories are important because people turn to them to understand what is happening in times of lasting stress; they help us appreciate other perspectives, develop empathy, and make sense of the world around us. In times of social distancing, stories provide a much-needed connecting platform where people may bond with each other, and where the shapers of policy can feel the effect of their decision-making in the lives of those around them.

As someone involved in teaching law at a public university in a developing country like India, my story will also be related to that world. Some of the elements in these stories can be found in other stories, from the lives of other people, while others may be singular to legal education in a developing nation.

The onset of the crisis

The story features three students and one teacher at a public law university in India. As the pandemic made its presence fully felt in the country, the government announced a nationwide lockdown on March 24, 2020, putting the

DOI: 10.4324/9781032690278-18
This Chapter has been made available Under a CC-BY-NC-ND license.

freedom of movement of almost 1.5 billion people on temporary hold. The announcement had taken our protagonists and the university administration by complete surprise. Most of the students were residing on the university campus, far away from their homes in other states, while the teacher used to commute daily from his home. The gravity of the situation and continuing threat the virus would pose for everyone over the coming months were yet to be apparent to anyone. The first priority was to send everyone back home before travel restrictions would come into full effect. Some students might even have felt a modicum of unexpected relief for what seemed like a short break amid a demanding semester and examinations looming ahead.

Rabiya's story

The first jarring note came when one of the students (let us call her Rabiya) realized she could not book a high-priced last-minute flight out of the city. A shy introvert, she kept debating whether to tell her friends about it. She knew that calling her family would not make any real difference. She was studying at this university for the last two years on an education loan while balancing an additional gig of teaching law aspirants to support personal expenses. Though her parents would want to, they could do little to pay the steep flight ticket price charged during this mad rush.

Rabiya considered staying back with one of her friends living in the same city. Surely this was going to be a short break of a week at the most! Then she remembered the scolding she received from her mother the last time she had to spend one night at her friend Reema's house, when she'd been staying with her mother's cousin during an internship in another city. She had only done so because Reema's house was within walking distance from the firm where they were both interning, and that time, they had dragged their tired selves out of work at 2 a.m.

So Rabiya booked a train ticket that never got confirmed and hoped she could manage the journey back somehow, even if it meant standing in an overcrowded train compartment for 16 hours. She realized that she could not carry all her study material with her under the circumstances, but she reasoned that with examinations almost two months away, there should be plenty of time for her to return and catch up on those readings. For now, her notebooks would have to suffice.

During the next 36 hours, she would be facing a harrowing journey back home, with the train station being mobbed as people desperately tried to catch the last train out. Several lives were lost in a mini-stampede, and the train made several unscheduled stops because the driver suffered from high fever and loss of breath. The laptop she had bought on a loan was damaged in the struggle, and she had to walk the last few miles due to the lockdown being in full effect by the time the train could reach anywhere near her destination.

Arghya's story

Arghya is the batch-topper in his penultimate year of law school, winner of moot-court competitions and debating tournaments, class representative of his batch, and favorite nominee in the electoral race for the university student body president. He was due to leave for an internship in a top UK Magic Circle law firm and would be applying for the prestigious Rhodes scholarship in a few months. Residing in the same city, he did not take the lockdown very seriously at first. In his mind, he was going through the features to highlight in his forthcoming email to the Supreme Court judge, whom he'd be serving as a research assistant to the next summer.

When he reached home, he found the place empty. A message left behind by his mother revealed that his father had fallen ill at work. A call placed to her told him little else, other than that several hospitals had no bed available for his father, whose condition was growing worse, that he required immediate oxygen supply and life support. After frantically calling everyone they knew and spending half of their savings over the next few days, they could get him admitted to a hospital, only to have him succumb to the virus within a week. The entire experience has left Arghya feeling bitter, helpless, and angry, his normally cheerful countenance and ever-helpful demeanor in marked abeyance.

Mathew's story

Mathew had joined law school less than a year back. His visual and partial locomotor disability notwithstanding, he had always been determined to study law. The possibilities inherent in a career facilitating access to rights and justice had excited him. He was looking forward to spending five years at one of the best universities in the country and starting community and grassroot-level advocacy upon graduation. Mathew's parents have been supportive of his choices and even shifted to this city when he joined this university to look after him better.

Coming to class every day means some painful moments for Mathew, especially when the university elevator breaks down and he has to be carried by his friends on his wheelchair up the stairs. It has been an uphill battle to convince the university administration to provide soft copies of the study material to him, as well as remind some of the teachers to explain their illustrations to him verbally every time they rely on PowerPoint presentations and the white board in the classroom. While Mathew's parents are proud of his resolve to soldier through these obstacles, the pandemic makes them worried, given the implications for his already weak immune system.

Rahul's story

The teacher, Rahul, is our fourth protagonist. He is a young alumnus of the same institution, who returned from industry to academia out of a sense of idealism and commitment to his own alma mater. He admits to loving his students as part of his extended family and also to his choice to stay close to his aged parents in the same city. When the university announced the suspension of physical classes, he found it irksome because his classes got disrupted. Yet he deemed this a temporary nuisance at most and even told his students that they should not treat this break as an excuse to neglect their studies. He did convince the university administration to give a week off to the students before online classes began, considering that the sudden journey back home at such short notice would doubtless make their lives more difficult.

Rahul also felt that the week would be crucial in providing teachers with some breathing space to figure out how to best shift from physical to virtual classes in as seamless a manner as possible. He had offered a few stand-alone lectures earlier to smaller groups using virtual platforms like Skype, but instinct told him that this might not prove to be so easy. For one thing, most of the older faculty members would not be comfortable with using any virtual platform. Setting up virtual classrooms, coordinating with students solely via emails and phone, and even having sufficiently strong internet connectivity to take hour-long classes comprising audiovisual feedback from many students – these were challenges that would need training to overcome.

How would the teacher react upon realizing he could no longer count on observing student body language in class to discern the extent to which they were able to understand the ongoing discussions? Would the absence of white board for illustrations or notes render the traditional pedagogy ineffective? When the students could not even see or hear each other clearly throughout the classes, nor engage in meaningful discussions and moderated arguments, could the Socratic method of teaching law survive such disruption?

Technology: purveyor of solutions and challenges

Rahul found most of these doubts come true over time. It was not that technology-based solutions did not exist for some of the problems the teachers and students were facing. As the pandemic continued, many products were on offer, ranging from online breakout rooms to facilitate mid-lecture brainstorming, to built-in recording options that allow asynchronous learning, integrated PowerPoint presentations within virtual platforms, virtual whiteboards and doodles for illustrations during classes, teaching workload management software, AI-based proctoring options, etc.

These solutions were not without accompanying challenges. Lack of access to devices, including computers, smartphones, and affordable data, posed a major obstacle not only for some of the students. It also affected some of Rahul's colleagues, both teaching and administrative staff, who resided in remote locations and were confined within their homes owing to the pandemic-induced lockdown, travel restrictions, and health risks. Even when access was possible, a lack of training prevented many users from using the devices effectively.

Dwindling student participation in virtual classes

Adequate student engagement during online classes proved to be a common problem. Rahul has always found his classes full of vibrant, intelligent, and eager-to-learn students, which to him represented the most attractive feature of his university. Yet participation in online classes displayed a dwindling trend. Moreover, it was difficult to identify the students who were trying to participate in virtual classes but could not do so for poor access. An example of this is Rabiya, who could not even get her damaged laptop repaired and had to use her aging feature phone to dial-in for most sessions. Together with sporadic local internet connectivity, this made even a sincere student like her miss many classes.

At the same time, even with access to technology, Arghya was finding it difficult to concentrate during these online classes. The trauma of losing a parent, the prolonged screen exposure, the stress of additional responsibilities at home after his father's demise, the pressure of being confined and deprived of his usually thriving and outgoing social life for weeks, and the anxiety of losing out on foreign internships and opportunities for further studies – the cumulative effect was starting to get to him with every passing day.

Rahul found that Arghya was starting to lag behind in academics, yet he found it difficult to discuss the matter freely with him in the way informal physical interaction might otherwise have allowed. On the other hand, he was amazed to learn from Mathew that to him, the online mode of delivery and the ability to learn from home have proven to be a blessing, as has been the policy decision by the university to provide mandatory digital course content to all students.

Pedagogy and learning challenges

One of the biggest challenges Rahul faced was finding the best way to modify and customize the existing course content, curriculum, and pedagogy to better fit the online delivery mode. Clinical legal education, deemed important in modern legal education, was getting affected by situational constraints. It was proving more difficult to connect the theories discussed in classroom during online classes with real-life case scenarios, exercises involving problem-solving, client counseling, mediation and negotiation exercises, and coordinated group discussions.

The practical exposure that students would otherwise get during internships was also being severely restricted during virtual internships. Such internships not only prevented the interns from developing skills associated with teamwork and group synergy, but they also limited effective mentorship by experienced seniors at work. The combined effect of these left an adverse impact on the university's and industry's ability to educate and train students to become the best possible legal professionals.

The allegations often leveled (occasionally with merit) at the lack of sanctity of online examinations conducted during this period, pointing out students' and the institution's inclination to compromise with learning and assessment quality, exacerbated the complications further. At times, it was assisted by overarching and ill-thought directives from educational regulatory bodies like the University Grants Commission.

Nor were resistance to change and inertia in favor of status quo helping matters. Rahul found this out to his chagrin when he tried to convince his colleagues to join a global group of academics trying to share experiences and brainstorm together in a hive-mind format to find practical solutions to the pedagogical and curricular challenges posed by the pandemic. It only received a lukewarm response, as did his initiatives to hold small, informal studying and mentoring sessions involving faculty and students. These discussions, sharing stories and doubts, could have helped all participants find a safe space, cope with mental health challenges better, and regenerate the institutional bond and community feeling that the pandemic and distancing have robbed.

Rays of hope

Yet all these efforts did not end in futility. In any place of learning, students are often the ones most instinctively attuned to the teacher's efforts. Rahul found out that while it might take other teachers longer to participate in such initiatives, a growing number of students were evincing progressively greater interest in those. Adversity often brings out the inner creativity of human beings, and when the intelligence, enthusiasm, imagination, and energy of students are guided by the experience and understanding of the teacher, the result may far surpass our dreams. The students seemed interested in pedagogical experiments involving participatory components compared to solo lecture sessions, especially when such experiments consisted of smaller groups.

Some positive results

At the same time, as the world and the legal education sector started absorbing the shock of the stimulus introduced by the pandemic, some of the positive results started to show. Rahul found that his colleagues' and students' as well as his own participation in global events, including conferences and competitions, have increased manifold during the pandemic. This was because of the possibility of virtual participation that was not only cost-effective (otherwise a matter of concern to students and faculty from developing economies that could avail little to no institutional financial support) but also time-efficient (previously, participation in such events might involve taking leave for a week, but virtual participation reduced such leaves to a single day, thus eliciting greater cooperation from institutional authorities). Of course, virtual participation reduced the possibilities of academic or professional networking to some extent. Yet, allowing such events to be conducted in a hybrid manner, even in the post-pandemic world, may produce better and more diverse academic output in the long run.

Similarly, digitization of lectures, course content, and virtual student and faculty exchange programs during the pandemic have allowed students and academics from every corner of the world to avail the benefits of global scholarship.

Conclusion and suggestions: the way forward

Increased use of technology in classroom and pedagogy (provided access can be ensured), continued training of educators in such technology, focus on open-book evaluation models consisting of practical application-based questions and assignments rather than rote learning. These are some of the changes introduced by the pandemic in Indian legal education, which should ideally survive during the aftermath of the pandemic. Individuals and certain institutions can and will adopt some or all of these practices, learning from their experiences.

However, for these changes to be sustainable and benefit the masses, the government needs to play a key role in framing the regulations and standards that would reflect such changes. It is also needed to implement such practices in all state-supported law universities via regulatory oversight through the Bar Council of India and the University Grants Commission. Digitization of all study material, sharing such material on open platforms for greater accessibility, provision of mental healthcare support to all students and staff, and similar other practices can also be considered as broader mandates for these institutions.

Principles and values

Above all, the principles and values of empathy, expression, cooperation, and adaptability should continue to be integrated as part of the education system. What the adversity of the pandemic triggered, let not the complacence of restoration of normalcy torn asunder. Today, more positivity is noted. Rahul looks at Rabiya leading movements in her state to provide legal assistance to homeless and migrant workers to cope with the pandemic. Arghya has set his own grief aside and is leading the student body by ushering in initiatives such as the procurement of smartphones and internet connection for needy students. He also heads discussions with the university authority about assisting students suffering from personal and financial losses during the pandemic. Mathew heads training programs for students, faculty, and non-teaching staff to help them understand and utilize new technological developments. He has gained confidence in knowing that the fate and future of Indian legal education are secure.

Although the pandemic may have ushered in many tragic changes, the challenges it has thrown saw Rabiya turn into a confident leader, Arghya into a greater empath and inspirer, and Mathew into a person who has grown in his creative strengths. That is why this story is not only that of Rahul, nor only of Rabiya, Arghya, or Mathew. It is of every student and every teacher, of those who are harbingers of change and believers in hope, resilience, empathy, expression, and education.

16
CHANGE AND CONTINUITY

COVID-19 and the Philippine legal system

Antonio G.M. La Viña

Introduction

The Philippine government under the Duterte administration fit squarely into the "strong man" category. Thus, it was no surprise and exactly as expected when its approach during the pandemic was of the same kind. The government "fought" the pandemic the same way it fought the drug war – through brute force. Over the course of the global health crisis and through several levels of classifications and alerts, the Filipino public knew one thing – the country was shutting down physically to prevent the spread of the virus. Less sure was what other measures were being taken to strengthen the already overloaded healthcare system or provide relief for the damage inevitable from these shutdowns. Least known of all was what the post-pandemic future would look like for the country.

A nervous public

"Lockdown" was the buzzword of 2020, and each time it was spoken by Malacañang, a nervous public would scramble to find out what was about to change. (Malacañan Palace is the official residence and principal workplace of the president of the Philippines.) Shutdowns came for transportation modes, mass gatherings, office places, schools – essentially all other major aspects of Filipino life. We had to avoid the virus and simultaneously adapt to a new normal overnight.

However, looking now with a retrospective gaze at the government's response, we can see that the strongman approach may not have been the best. Through the lens of the law, we must review what was attempted, what

DOI: 10.4324/9781032690278-19
This Chapter has been made available Under a CC-BY-NC-ND license.

was experienced, and what lingers on in Filipino society. Was the law used efficiently, effectively, and empathetically? Was the government's hard stance effective or a failure?

Law used as a tool by the government

The government employed the tools of the Constitution to justify its strongman approach. On the other hand, the law was used responsively, almost by reflex. There was a sore lack of foresight and forward thinking. Instead, it felt like policy was wielded in support of the trampling over of ways of life and significant loss of livelihood, and curtailed personal freedom. As we will see later on, the legislation enacted in response to the pandemic lacked a key feature – recovery. Reflexive legislation acts as a band-aid, but effective legislation foresees future problems and strengthens the legal system to capacitate it to act and not react when these occur. From this, we can see that the legal system of the Philippines was used to perpetuate the same types of hardline policies the Duterte government was known for, rather than as an actor of reform in the face of an unprecedented emergency.

Of course, some aspects of the pandemic were entirely unpredictable. No country and no legal system in the world, for that matter, possessed all the tools to enable an efficient and effective response to the pandemic. But there were frameworks that could have guided better, quicker reforms, such as the Health Emergencies Program of the World Health Organization (WHO) that were available as the basis for national emergency response.

Yet the Philippines today is an example of what happens when everything that can go wrong goes wrong. The national economy is struggling to return to its prior positive projections, inflation rates continue to climb, and many sectors of business have slowed as they adjusted to the new normal. Socially, the damage caused by the world's longest hard lockdowns is impossible to quantify yet, as the consequences will surely be felt for years to come. All of these negative effects came in pursuit of a safety that was not wholly achieved.

The Philippine legal system

Article III of the Philippine Constitution is the Bill of Rights, enshrining essentially the individual freedoms we are each guaranteed. Critically, it establishes the limits of the State's exercise of its lawful powers. The Bill of Rights protects the individual from abuse of the State by guaranteeing civil and political rights that stand strong in the face of the awesome powers of government.

However, all rights are not absolute, and the State is permitted to lawfully limit these for the greater good. An apt example is that Article III also gives the State the power to limit the right to travel if doing so would be in the interest of public health. The liberty of the individual and the power of the

State are meant to sit in balance. Thus, was it truly within the scope of powers of the national government to curtail personal freedom to the extent it did during the lockdowns? Were the interests of the public served to justify this overstep of rights? The pandemic was a truly unprecedented event that warranted extraordinary action – but did the Filipinos instead get extraordinary force? The government's Covid strategy may have been overprotective, or it may just have likely been simply overzealous and underplanned.

Two central pieces of legislation

Two key pieces of legislation were passed by the national government to address the social and economic aspects of the pandemic. For one, Republic Act 11469, or the Bayanihan: We Heal as One Act, declared officially the national health emergency of COVID-19 throughout the Philippines (Dominguez, 2020). (Bayanihan means working together as a Filipino community to achieve a common goal.) The same bill authorized the president to exercise special powers, limited temporally and bound by certain restrictions. The serious health threats posed by the pandemic virus and the disruptions that were necessitated by the same were the basis of the law, though the legislative act came after measures had been already taken by the president against the situation. States of public health emergency and of calamity were already declared by the chief executive through Proclamations 922 and 929, respectively, before the bill passed the Houses of Congress. More significantly, areas throughout Luzon had already felt the force of lockdowns.

Bayanihan One

Bayanihan One was the first time that a national health emergency legislated emergency powers for the president's exercise under the current 1987 Constitution. But it was reactive, not responsive. Adapting, not anticipating. The Philippine response to the pandemic was marked by these issues. The law used the tools of the legal system to provide the executive with additional powers, in theory allowing for a quicker response. Ideally, these extraordinary powers would have helped the government stay one step ahead of the virus. But this ideal was not achieved. In actuality, the executive used it merely to strengthen the already lockdown approach in place.

Bayanihan Two

Following Bayanihan One was Bayanihan Two, or the Bayanihan: To Recover as One Act (Congress of the Philippines, 2020). This time, the law was employed in a way to address the rising crop of socioeconomic issues given rise by the pandemic. This act focused on social ameliorations and socioeconomic placation. In a desperate bid to hold off the teeth of poverty biting at the heels of Filipinos who suffered loss of livelihoods during the lockdowns, as well as a means to somewhat stimulate the economy by restoring through very limited capacities the buying powers of Filipinos, the bill provided loan payment easements and provisions for government subsidies.

The special powers granted to the president did away with some typical procedural delays and allowed for quicker implementation of threat mitigation policies. But the strategies continued to be lockdowns, shutdowns, and quarantines.

Moreover, the Bayanihan Acts were nothing novel. Every country rolled out some form of social amelioration, some form of unprecedented national response to try and stem the flow of the virus. The Philippines legal system is capable of quick action through cooperation between its branches of government. Congress can call for special sessions and pass legislation in a manner that properly and legally bypasses the traditional process to enable the executive branch to quickly and effectively adapt to the ever-changing Covid landscape. Yet while Bayanihan One was passed quickly, these powers were not exercised to the extent they could have been, and the country instead fell back to the strongman approach. Although Bayanihan Two was meant to provide mechanisms to accelerate recovery and bolster existing systems, the present situation of the country clearly shows that this too was a failure.

"No vaccine, no ride" policy

The conflict between the legal system and the national Covid response is encapsulated well in the "no vaccine, no ride" policy that was attempted for implementation in 2021. President Duterte, through the Department of Transportation, sought to implement a policy that banned all non-fully vaccinated persons from riding public transportation. There was no legal basis for this exercise of State power. In fact, the order was questioned by legal scholars who correctly pointed out that it was an encroachment on the powers of the legislative branch and a direct contradiction to existing laws. In its pursuit to continue its policy program, the government overstepped its bounds and created more problems for Filipinos.

For context, when the controversy arose in January 2021, only around half of the Philippines' 120 million population had been inoculated with one dose. While the policy was only rolled out in the capital region before being withdrawn after immediate backlash from the public and opposition politicians, the ill-thought-out policy still caused immeasurable economic and social damage for the brief moment it was around. If it had stuck around and rolled out nationwide, nearly half of the nation would have been banned from public transportation for not being fully vaccinated. Funnily, they would have no way to go to vaccination sites.

The Philippine legal framework

In short, the Philippine legal framework was an active actor in the national Covid response, though not in the solution. Rather, it helped answer the repeatedly asked question: Is this a valid exercise of State power? How important are human rights in the time of the pandemic to the government?

As the threat was a global one, comparisons are easily made between the approaches of different countries. It should be remembered that while today, the idea of a nationwide lockdown seems well within the realm of possibility, prior to 2020 it was almost incomprehensible. Worldwide, unprecedented measures were implemented by nations as they struggled to stem the spread of COVID-19 through any means necessary. The strategies designed were varied and so were the results.

Legally, can we say that the powers enshrined in our Constitution were wisely used? Were the democratic processes designed to empower every Filipino responsive, reactive, and relevant to the needs of the common man? Even with several rounds of legislation giving measly amounts of social amelioration, loan extensions, and calling for national responses to the crisis, was the Philippines truly equipped with the tools to adapt and overcome the challenges of the virus? The results speak for themselves.

Now, let's delve further into the actual effectivity of the approach taken by the Philippine national government, the flaws, the falters, the successes, and the outcomes. Granted, there is some merit to an authoritarian approach. If successful at preventing large outbreaks, it could provide some relief to overburdened public health systems such as in the Philippines, where rural healthcare penetration has been a persistent issue. The authoritarian approach, at its heart, was a preventive one.

Failure to adapt

Yet with each new strain and surge, the Philippines failed to adapt. As the pandemic dragged on, the health crisis was joined by social and economic issues that were natural consequences of the lockdowns. By attempting to stem the spread of the virus through physical policies, the Philippines crippled itself in other ways. One illustrative example is found in the tourism sector: Normally contributing more than one-fifth of the national GDP prior to the pandemic, it dropped to just 4.8 percent in 2020. By 2021, this recovered to just shy of over 10 percent, but projections by the World Travel and Tourism Council peg recovery of over 20 percent to be hit in 2032 (WTTC, 2022). How many families and communities will have to struggle to adapt in the decade before the industry bounces back? And that's just one industry. The lockdown approach anticipated its own success, to the point where it didn't include a recovery plan.

The Philippines, rather than using the time bought by the lockdowns to strategize its next moves, had just sat and hoped no more fires would start. But fires started and spread.

Global comparisons

For a virus such as COVID-19, exercising the government's strong arm to curtail the freedom to travel and gather can save lives, as evidenced by other hard-stance countries like Vietnam and Japan. These nations locked down hard and quick, preventing significant loss of life. However, an authoritarian take on a health crisis can only work if coupled with responsive action on emerging social issues. Lockdowns also need to be married to efficient government action in controlling outbreaks, employing tracing mechanisms, and planning for vaccine deployment.

Looking at countries that took an opposing approach such as the United States and Brazil, which were much more liberal in their dealings with the virus, we see the other side. Economic bounce back occurred soon after one of the largest declines in recent history to financial markets. Surely, there are criticisms that can be leveled against these nations and their approaches, specifically for loss of life, but it is undeniable that their national economies were able to weather and wean off the effects of the pandemic whereas the Philippines remains stuck in the sticks.

Outside the economic aspect, many countries that relied instead on broad and free testing and contact-tracing systems were able to avoid total shutdowns while still minimizing human costs.

The Dengvaxia scandal

It bears noting, however, that vaccine hesitancy and difficulties securing vaccine supplies were a uniquely Philippine issue due to the prior Dengvaxia scandal. For those unfamiliar, Dengvaxia was an anti-dengue vaccine manufactured by Sanofi Pasteur that was deployed by the Philippines until controversy arose in 2017. The Philippines was the first Asian country to approve the vaccine for commercial sale. Between April 2016 and November 2017, over 700,000 individuals were given at least one dose through mass inoculation efforts of the Philippine Department of Health. The vaccination program was specifically targeted toward children and drives were done in primary education institutions.

The controversy began when in November 2017, Sanofi announced that further analysis of their vaccine found that those who had previously not been infected by the dengue virus, of which there are four strains, may actually suffer more severe incidences of disease following vaccination if they become infected. At this point, the Philippine government had given thousands of children doses of the vaccine without having this critical information of long-term effects. Sanofi was then aggressively pursued by the Philippine government for accountability.

The mistakes of Dengvaxia haunted the Duterte administration's rollout of COVID-19 vaccines. There was, of course, public hesitation, but there were also concerns for the government as to the liability of another botched inoculation attempt. The Philippines was the last Southeast nation to receive its initial doses of vaccines. These only came into the country following the passing in February 2021 of a law that gave indemnity to vaccine makers in case their COVID-19 vaccines caused adverse side effects (Reuters, 2021).

After the scandal of Dengvaxia and the Philippines' attempts to hold Sanofi accountable, vaccine manufacturers feared to be placed in the same situation. They tied the hands of the Philippine government to guarantee them immunity from suits before any vaccines would be delivered. While the Philippine government debated this rock-and-hard-place dilemma, its neighboring countries began their mass vaccination programs. After the passing of the law, the Philippines began its slow trek into the new normal with an end somewhat in sight.

The new normal: changed for better or worse?

Now we are out of the tunnel, it may seem. We were looking back, though many still feel as if the pandemic never left. The new normal is here. In the Philippines, especially, the poor government response has led to the impossibility of a return to what was before. COVID-19 was a time of incredible flux across the world. Nothing was the same. Nearly all systems of human life were forced into the choice: change or continue?

For the Philippine legal system and the national government's approach to the pandemic, the choice was continuity. The same old approaches to new and novel problems proved disastrous in the end. The Philippines ended up in a worse place economically and socially because of the pandemic. Poised to be a developing country in 2019, it now lags further behind its neighbors and struggles to catch up. While the economy drags its feet and begins to recover as Covid restrictions are lifted, ordinary citizens continue to suffer the consequences. Inflation as of August 2022 was 6.3 percent following five months of consecutive acceleration (PSA, 2022).

Thousands of lives may have been saved from contracting the virus, but millions instead were destroyed. The statistics of COVID-19's impact in the Philippines need to count the livelihoods lost, the economic damage to come from children being educated online for years on Southeast Asia's least affordable internet, the mental toll, and the financial strain of rising goods. The nation's vulnerable were made more so because of the ham-fisted lockdown approach. Social challenges and systemic issues were exacerbated, and the government failed to keep up with the problem.

The state of the new normal

Undoubtedly, the pandemic has left in its wake an atmosphere of radical change. The new normal is indeed a full departure from what came before, separating by temporal division what was and what is now.

Education and healthcare are two fields that show these radical shifts to a high degree. Only in 2022 have children returned to face-to-face classes on a staggered schedule. Rather than change to find some solution over the long years, the Philippines continued with its hardline approach long after other countries saw the potential damage it was doing. What are the consequences for this generation that learned online through their formative years on pricey and unreliable internet connections? How can we quantify the losses and damage to this generation of children? Beyond the economic losses of projected productivity, how can we quantify the loss of community, of learning environments, and of culture?

In sum, the draconian approach taken by the Philippine government has given all of the negative effects and none of the possible positives. Despite the lockdowns, the Philippines still got hit by wave after wave of cases. Surges filled hospitals. Faced with harrowing triage decisions, doctors had to make choices based on bed availability. Patients with other diseases like cancer had to defer treatment and checkups as hospitals were simply overwhelmed. Mass testing and vaccination were both deployed late. Filipinos essentially sat at home twiddling their thumbs, cowed by the cloud of a virus that could decimate their families, and without a solid plan from the government.

COVID-19 brought on changes that we can never undo. The new normal is reality. The Philippines learned the hard way that if you cannot adapt or if you refuse to adjust, you will sink. Now, we wade through the waters of an uncertain future, pulled down by the weight of what has been lost.

Reference list

Congress of the Philippines. (2020, July 27). *Republic Act No. 11494: An act providing for Covid-19 response and recovery interventions and providing mechanisms to accelerate the recovery and bolster the resiliency of the Philippine economy, providing funds therefor, and for other purposes.* www.officialgazette.gov.ph/downloads/2020/09sep/20200911-RA-11494-RRD.pdf

Dominguez, C. G. (2020, April 1). *Implementing rules and regulations of Section 40 of the Republic Act No. 11469, otherwise known as the "Bayanihan to heal as one act".* www.officialgazette.gov.ph/downloads/2020/03mar/20200401-IRR-RA-11469-RRD.pdf

Reuters. (2021, February 26). *Philippines' Duterte signs indemnity bill for COVID-19 vaccine rollout.* www.reuters.com/article/us-health-coronavirus-philippines-idUSKBN2AQ268

Philippine Statistics Authority. (2022). *Summary inflation report consumer price index (2018=100): August 2022.* https://psa.gov.ph/price-indices/cpi-ir/node/168112

World Travel & Tourism Council. (2022, April 20). *WTTC's latest economic impact report reveals significant recovery in the Philippines Travel & Tourism sector in 2021* [Press release]. https://wttc.org/News-Article/WTTCs-latest-Economic-Impact-Report-reveals-significant-recovery-in-the-Philippines-Travel-and-Tourism-sector-in-2021#:~:text=Before%20the%20pandemic%2C%20the%20Philippines,share%20towards%20the%20country's%20GDP

17
DIGITAL TECHNOLOGY
A best friend for implementing COVID-19 policy in China

Li Xudong

Introduction

At the beginning of 2020, the COVID-19 pandemic first struck Wuhan, China, causing thousands of people to get affected. Every single hospital was over its full capacity. Doctors and nurses in Wuhan worked extra hours day and night but still could not manage the influx of patients. This led doctors and nurses from all over the country to flood into Wuhan to tackle the pandemic. However, since we did not have vaccinations or proper treatment for this new disease, we still lost a lot of patients.

In order to avoid the pandemic from getting worse, China designed and carried out a highly efficient response system to tackle it, which included major lockdowns, mass medical nucleic acid testing (a medical test for identifying positive COVID-19 cases), strict surveillance and isolation of virus-infected people, ongoing quarantines in areas with positive infection cases, and continuous border control. Although the policy may sound brutal, it is credited for controlling the fatality rate at a remarkably low level, keeping the COVID-19 virus within China's border as fast as it could, and more importantly, enabling residents to work and live in a comparatively safe environment.

As of September 2022, even with other countries' announcements of dropping COVID-19 policies and trying to go back to the norm, China still insists on its Zero-Covid Policy. This means the virus should be eliminated locally, and in the end, the area should resume normal economic and social activities.

To achieve this grand goal, digital technology has been playing an increasingly important role. The three most used digital technology products in managing the pandemic are the Health QR Code, Travel Card, and Venue QR Code. I am going to discuss how technology is helping carry out the Covid

DOI: 10.4324/9781032690278-20
This Chapter has been made available Under a CC-BY-NC-ND license.

policy in China. I will also provide personal experience as one of the citizens in Shenzhen that went through major lockdowns.

The Health QR Code

The Health QR Code is a locating and tracking application used as an electronic pass certificate for individuals during the COVID-19 pandemic in China (see Figure 17.1). It shows the health status of the certificate holder and is used to enter and exit places where the certificate is required. People that apply for the code are required to fill in basic information, including personal identification, travel history, place of residence, possible symptoms, current body temperature, and whether the applicants have been in contact with suspected or confirmed COVID-19 virus-affected patients.

FIGURE 17.1 Screenshot of Li Xudong's Health QR Code with English explanation.
Source: © Li Xudong

Based on the information they reported and the big data of their travel history obtained by the government through digital tracking, applicants of different risks are given different health code levels – green, yellow, and red. A green health code means the lowest risk of spreading the virus. It indicates that the code holder does not show any related symptoms and has no travel history in high-risk areas or contact with high-risk individuals. A neutral yellow health code means that the applicant has some potential for spreading the virus because the person had close contact with the infected or because the area that the individual lives in has a certain number of positive cases. Apart from this, people from outside this country are automatically given a yellow code, which will turn green after self-quarantine for days and a negative nucleic acid testing result.

A red one, however, is much more serious. Red means the code holder is suspected or confirmed to be a case of COVID-19 virus infection. This also includes asymptomatic patients. All relevant health information of people will be updated to a national big data network. People will receive messages or calls that remind them to take specific actions based on their Health QR Code levels, like taking immediate tests, self-isolation, reporting to the neighborhood that people live in, etc.

Different codes have different meanings and affect code holders in multiple ways. Green code holders are free to travel as long as they provide their code to onsite security. Normally, they do not need to be quarantined when they arrive at a new place. However, yellow and red code holders are much more restricted. Yellow code holders are not allowed to travel and must go through self-quarantine in their own places of residence for several days before they are granted a green code. Red code holders are required to undergo intensive medical isolation in specific places arranged by the local government.

At the beginning of the pandemic, housing and food are free of charge, but now people in isolation have to pay a certain amount of money that equals the market price of living at the time. Moreover, both yellow and red code holders must take multiple nucleic acid tests for negative results, and in some extreme cases, the tests are done on a daily basis.

Shenzhen

When I was in Shenzhen, I woke up one day to an announcement that lockdown measures and mass testing would be conducted throughout the whole city for at least seven days. During this lockdown, every resident was required to take at least three nucleic acid tests in the temporary testing sites built in each neighborhood. Luckily, the tests were all free of charge at the time. However, there were other places where people were required to take regular tests and pay for them, usually at a very affordable price of less than USD 1. With the three-level health code system, testing was conducted in a very efficient way. Medical staff only needed to scan the code to get all relevant information, including vaccinations, the last testing results, and the health status of the test takers. If residents prepared the code in advance, it took no longer than one minute to finish the testing.

For yellow and red code holders, they were taken to specific testing sites separated from the green ones. I was lucky that I had a green code during the entire quarantine, so I stayed in my neighborhood for a week before I became a free man again. Ever since then, the government of Shenzhen required its citizens to take regular tests to go to work or attend other social events. Normally my colleagues and I were asked to keep a valid test result from a test conducted within 48 hours. The requirement can be adjusted to 24 hours, 72 hours, and other times due to the seriousness of the pandemic. Every morning, the security of my work building would ask us to scan a code to pull up our most recent test so we could enter the building.

Xining

I experienced getting tested regularly in a different city when I went to Xining to visit my mother in July 2022. Compared to Shenzhen, Xining is a smaller and less developed city situated in the northwest of China, with comparatively less perfect infrastructures and a less wealthy government. Thus, the testing sites in Xining were much more sporadically scattered in the city, and people usually spent more time traveling to the places to get tested and waited in line for longer since the staff was also limited. I stayed there for a whole week and got tested three times. It took me and my mother about half an hour to walk to a testing site. I once saw a kid in a Superman costume waiting in line for a test, which I thought was quite amusing (see Figure 17.2).

Shanghai

The topic of what people from different places wore when they went out to do the test had been viral on Chinese social media apps. The most worth mentioning was that people from Shanghai tend to dress interestingly, such as suit-and-ties, nightdresses, cartoon costumes, etc. People entertained themselves by treating the tests as runway shows, since it was the only reason they went out during quarantine.

The Travel Card

Another good friend in implementing the COVID-19 policy in China is the Communication Big Data Travel Card (Travel Card). This is not an actual card, but a software developed by the China Academy of Information and Communication Technology (CAICT) (Triolo & Webster, 2018).

Digital technology 153

FIGURE 17.2 A kid in Superman costume lining up for mass testing in a residential area, China.

Source: © Li Xudong

Using cell phone signaling and Bluetooth positioning, this app can track the places that phone users have been to in the last 14 days (starting from July 8, 2022, it is limited to the past seven days) and mark places that have confirmed cases.

A grading system

In order to make this travel card work, the Chinese government invented a risk grading system to evaluate the level of risk of the cities in China. It divided areas into three categories: low-risk areas, medium-risk areas, and high-risk areas. Medium-risk areas are places that confirmed cases or asymptomatic infected patients visited or stayed in. High-risk areas are usually the place of residence of confirmed cases or asymptomatic patients. The specific criterion in categorizing different areas is based on the discretion of each local government. For example, if a person went to a medium- or high-risk area, the name of that city will appear on his or her travel card with an asterisk beside it. (Before August 20, 2021, the app used to mark the names of the cities with medium and high risk in red.)

With the three-level category of risk, each local government can easily enact different traffic restriction policies on people with different travel histories. People who have travel cards without any medium- or high-risk areas may travel freely within the country. However, those who have asterisk indications on their travel cards might need to provide recent negative nucleic acid test results before traveling and undergo self-quarantine or isolation after arriving at their destination. In some circumstances, the local government might even refuse to let people with marked travel cards to enter the city. With the travel card system, every local government can impose personal-tailored travel restriction policies on travelers with different risks, controlling population movement throughout the country and preventing the virus from spreading nationwide.

Utilizing digital technology

I also experienced the convenience of utilizing digital technology in managing people's movement by local governments. From January to June 2022, I was an intern at a law firm in Shenzhen. As I was constantly on business trips, traveling was somehow unavoidable for me at that time. I remember when I traveled back to Chongqing from Shenzhen, my travel card showed an asterisk on Shenzhen because at that time, Shenzhen was one of the medium-risk areas reported to have a certain number of positive cases.

When I arrived at Chongqing Airport, I was told to scan a QR code on which I could report my travel history and health status on my cellphone. It was so convenient that it took me less than three minutes to finish. After that, I received a message from the local government saying that I needed to take two nucleic acid tests within three days, with the interval needed to be more than 24 hours. Apart from that, all the test fees were paid by the local government, and the tests could be easily conducted in one of many temporary testing sites built in the neighborhood where I lived in.

The Venue QR Code

The Chinese government not only manages the pandemic through each individual's travel and health data, but it also requires every store, restaurant, and other public places to generate their own venue codes for collecting data of all visitors. People who need to enter a public place have to scan its venue code. After that, the Health QR code of each visitor would automatically appear on his or her cell phone for security to inspect. If the QR code was yellow or red, the visitor would be denied entry.

All visitors would be kept in the national big data network. If positive cases were ever reported from such a place, the government would notify all visitors that might have been infected. With the small electronic devices at people's palms, security could instantly know the health status of visitors and the government could keep records of potential spreading of the virus for following preventive actions. Again, all of this could not be done without the help of electronic technology.

Conclusion

It has been over two years since the initial outbreak of the COVID-19 pandemic in China. The design and utilization of digital technology in managing the pandemic are improving rapidly to better adapt to changing circumstances. For example, with the aim of making the Health QR code apps more user-friendly to the old and young, nearly all these apps allow people to add and manage their family's health profile. This way, old people and children who are not that familiar with smart devices can go with their adult family members to take nucleic acid tests and attend other social events.

Since my mother is not a huge fan of smart electronic devices, I became the family member who has access to her Health QR Code. If she travels with me, she would not need to take out her own phone because I could simply pull out her profile right away. For people with disabilities, they could simply tap on the "Read" button to listen to, keep track of, and show others their health data.

Conducting public management

Since the outbreak of COVID-19, China has maintained a relatively strict pandemic policy. While residents cannot live as hassle-free as they would have if the pandemic did not exist, it is thanks to data technology that the government can provide as efficient and convenient a pandemic management policy as possible while still ensuring safety. The health codes, travel cards, and venue codes are excellent examples of technology that help the government conduct public management more easily and efficiently in this hard time.

As someone who has experienced multiple rounds of lockdown, traffic restrictions, and large-scale nucleic acid testing during the pandemic, I have a deep sense of the convenience that technology brings to us, which can shed light on the future direction of the development of public management technology and detailed implementation of public policy.

On the other hand, I have been thinking about the necessity of continuing the strict style of pandemic policy in China. Judging from what I see in the United States, where people tend to have a more laid-back and flexible attitude toward the pandemic, I think it may be fine for China to be a little bold now, too, especially since most of us have already been vaccinated, and the symptoms of this virus do not seem to be as terrible as they were when it first hit us. Moreover, having a less restricted policy would not only lead people to have a normal, more interactive, and engaging lifestyle, but also boost our economy. I bet most people miss the old days.

Reference list

Triolo, P., & Webster, G. (2018, October 16). *Profile: China academy for information and communications technology (CAICT)*. www.newamerica.org/cybersecurity-initiative/digichina/blog/profile-china-academy-information-and-communications-technology-caict/

18
THE GOLDFISH AND THE NET(WORK)

Nicole Mazurek

Introduction

At the time of writing, a number of government orders made in Australia during the pandemic are finally being lifted, including those that restricted gatherings and travel, imposed vaccination mandates and face mask rules, all of which limited people's abilities to live full lives. As we come to the end of this rollercoaster ride of government mandates and, as will be described later, self-imposed restrictions, the world is returning to normalcy and our vision is clearing. We can look back over the last few years and consider how well we handled the COVID-19 response.

The convoluted style of reporting by the government and social media on the pandemic contributed to growing anxiety in Australia. Additionally, government orders constantly changed and at times were difficult to interpret, with governmental guidance on how to interpret and implement them coming too late. Interestingly, there were decisions made entirely outside of the administrative and legislative spheres – decisions made by parents, business owners, and others – that caused harm that could have been avoided. This chapter will focus on this last aspect.

I come to realize that in comparison to others' experiences, I had very few personal struggles during the pandemic. I belonged to that group in society that disproportionately benefited from the societal changes the pandemic instigated – from the introduction of working from home arrangements across the legal industry to various investment opportunities. To quote a former colleague – without intending to be callous but merely highlighting the differences in experience – "Covid was great for us!" While the changes in society did not hurt me personally, they did affect me in some ways. Thus, I was certainly not blind to the adverse effects that these changes had on others.

DOI: 10.4324/9781032690278-21
This Chapter has been made available Under a CC-BY-NC-ND license.

News and social media

One may look resignedly at news and media outlets (whose ownership is highly concentrated, meaning there are few owners) that were the primary deliverers of information and updates about the status of the pandemic (Jolly, 2014). The style of news reporting was stress-inducing, to the extent that several of my friends and colleagues admitted with heavy hearts that they had vowed to stop watching and listening to news coverage. Gradually, they fell out of lockstep with the daily anxiety high and, as a result, came to feel more grounded, happy, and free from inputs weakening their mental states. If this stress-inducing style of reporting was intended to mobilize people into complying with government orders, it failed to achieve that objective by losing their engagement.

On one occasion, the government released an advertisement that was intended to be graphic and distressing, to motivate young adults to get vaccinated at a time when the vaccine was not yet widely available to those under the age of 40 (Murphy, 2021). Its only effect was to create a sense of hysteria and powerlessness.[1]

Children face the pandemic

I know of an eight-year-old child whose parents were so deeply alarmed by what they had heard about the dangers of the virus that they kept their daughter from going outside. For extended periods of time, their daughter was unable to socialize with her friends or have friends visit her. Guided by her parents' concern for her health, she was not allowed to play in the park, stand under the sky, or hear the sounds of the world other than the monotonous hum of apartment living. As her parents did not let her have contact with any of her friends, she resorted to having a goldfish as her friend.

The goldfish

Unfortunately, one day the goldfish died and, as the fish was being carried out in a net, something inside this child snapped. The stress that had been building up within her came to the fore, and she fell into deep desperation. Being a child, she was ill-equipped to handle what she felt and unleashed the frustrations she did not yet have the skills to manage – by screaming at her parents, saying how much she hated them and could not stand being around them anymore.

This short story was one of many tragedies the pandemic caused. It seemed that the parents' instinct to protect their child was dominated by solutions to keep her from the virus at the cost of all else. It was in fact the parents who were scared to death – paralyzed by fear. Although they were not exactly unliving like the goldfish, they were so stripped of their autonomy because of their fears that the result was much the same. Once they were swept up in a net of panic and hysteria, they, as well as their daughter, were carried away to whatever unhappy place it took them.

We are all sympathetic to the parents' concern for the health of their child, but would you not ask yourself if their response was proportionate when weighed against the adverse effects that such isolation could and did have on their daughter?

One reason this child was imprisoned in the apartment was because local governments in every state and territory in Australia mandated school closures, though the extent and period of closures varied across jurisdictions. In New South Wales, where this child resided, there were some exceptions where students in metropolitan and surrounding suburbs could physically attend school. One was if their parents or guardians were essential workers (such as doctors). Another was if the children were 12th year students and needed to be on campus for their higher school certificate preparation. However, other schoolchildren continued to learn from home remotely.[2]

A survey conducted by a university research team found that the mental well-being of students suffered significantly during this time. Many students described learning from home as one of significant stress, anxiety, and frustration (Gore et al., 2020). One does not, however, have to rely on an official survey. Ask a parent, teacher, doctor, or social worker to confirm their experiences with children's emotional well-being.[3]

Stay-at-home orders

On March 31, 2020, the government made an order that took on quite a severe character and was in place largely in this form until May 15, 2020.[4]

Referred to as "stay-at-home orders," this order defined permitted gatherings as a gathering of only two people in a public place. There were exceptions to this, such as a gathering for the purposes of work, a gathering to provide care or assistance to a vulnerable person, or a gathering of persons from the same household. Stay-at-home orders also directed that a person must not, without a "reasonable excuse," leave the person's place of residence. The order specified examples of reasonable excuses, like getting food or other goods and services; traveling for the purposes of work or education if it is not possible to do it at home; exercising; or for emergencies or compassionate reasons.

At first glance, one wondered whether this order achieved anything at all, as those were the main reasons most people left their homes for any event and, accordingly, the order seemed minimally disruptive. It appeared that unpermitted conduct was sitting in solitude outside on a bench for no apparent reason. However, despite how effectless the order seemed to be, its effects were in fact intensely palpable.

Social conversations began to echo exchanges in a courtroom – the speaker would recount their experience of walking in the park with friends on the weekend, sometimes supported by proof that their activities conformed to the prevailing government orders. Listeners would cross-examine the speaker whenever he or she appeared to have misunderstood the scope of the orders. Oftentimes, the speaker had simply misspoken and needed to, with apparent embarrassment, clarify the fact that he or she had misspoken and had, in fact, always complied with the orders.

Conversations would go something like this:

Defendant: This weekend I met with a few friends at the local park. It was very refreshing.
Examiner: I hope you social distanced! It's only permitted for two people to gather together in a public place (or more if you're from the same household)!
Defendant: Oh yes . . . I mean, no . . . of course! We just happened to run into each other. . . . I was just taking a walk with my partner. Of course we social distanced.

I also overhead a discussion between two lawyers where the junior lawyer said she had been visiting one of her parents who lived alone. The senior lawyer asked whether that was, for the purposes of the order, a compassionate reason. As any lawyer retained to answer that question would know, the term "compassionate reason" was not defined in any government order. Thus, the best way to understand what it meant would be for a judge to hand down a decision on that very question. However, by the time a judicial decision has been handed down, it would be too little too late for people to be able to adjust their conduct accordingly.

Ultimately, however, the meaning of the term "compassionate reasons" became irrelevant because the order was amended a month later. The new version stated it was a reasonable excuse to visit another person to provide care or support for the mental, physical, or emotional health or well-being of that person. But then, one wondered, what was a compassionate reason?

The point here is this: The government orders were living documents, so regularly changing them was certainly difficult, as was keeping abreast of each change, even if the interpretational issues could be resolved. This complexity surely contributed to the general anxiety the community felt, not least because individuals could not come to a definitive conclusion that was free of risk to them. The penalties for breaching the orders were not insignificant. If an individual failed to comply with an order, they would be risking either an on-the-spot fine of anywhere from AUD 40–1,000 (about USD 30–700), to the maximum penalty of imprisonment for six months or a fine of up to AUD 11,000 (about USD 7,500) – or both – plus a further AUD 5,500 (about

USD 3,800) fine each day the offense continued (Legal Aid NSW, 2022). Corporations that failed to comply with a direction were liable to a fine of AUD 55,000 (about USD 38,000) plus AUD 27,500 (about USD 19,000) each day the offense continued. In such circumstances, being uncertain about whether or not visiting one's relative was for a "compassionate reason" or whether leaving home to go somewhere else was permitted took on a significant degree of relevance.

In August 2021, similar stay-at-home orders were made, with the addition of a geographic limit for people living in Greater Sydney. The orders said individuals could only travel within their local government area or within a five-kilometers radius from their home. Similar to the orders of March 2020, these specified that it was a reasonable excuse to be away from one's place of residence to provide care or assistance "to a vulnerable person," or for compassionate reasons, including two people in a relationship but did not live together. This time we asked ourselves what makes a person "vulnerable" in the eyes of the order. And if it was "compassionate" to visit a romantic partner, was it also "compassionate" to visit a parent who lives alone? These kinds of inquiries eventually became quite tedious as a consequence of their indeterminateness.

Another palpable effect of the government orders was the inappropriately political character that discussions seemed to take on. On one occasion, I observed a conversation between an acquaintance and his work colleagues. To me, he had expressed his frustrations about the challenges of school closures, while to his colleagues, he expressed a diametrically opposite view, emphasizing the benefits to family life of schooling from home. I would have not thought much of this had it not been such a stark contrast. It appeared as though he had hesitated to express his frustrations in a professional environment, in case it would be perceived as criticism or opposition to the measures altogether. If this is the correct interpretation, it is indeed a very unfortunate one.

Data privacy

Other indicators suggest that a number of responses to the pandemic were rushed, which opened up the possibility of adverse consequences for individuals. One example relates to how companies and other entities handled proof of an individual's vaccination status. Government orders required certain groups, such as airport, education, and healthcare workers, to be vaccinated. Other entities not affected by these government orders also requested proof of vaccination status, in an attempt to manage the work, health, and safety risks of unvaccinated individuals entering business premises. Effectively one policy was implemented across businesses, despite the government only requiring such measures in certain cases.

In order to effectively implement these orders and internal policies, an entity needed some way to check that the individual met the vaccination requirements. One way to do this was to sight the relevant document – merely looking at it rather than writing down anything about it. The other was to "collect" it; under Australian privacy law, information is "collected" if it is included in a record (that includes keeping a copy or making a written record of the information in the document). The relevant document could be an individual's immunization history statement or a COVID-19 digital certificate available through the Medicare platform or a mobile wallet, or a document containing information exempting the individual from the vaccination requirements.

Not all state governments made orders that permitted the "collection" of vaccination status information. However, presumably many entities collected such evidence because it would have been impracticable to sight it on each occasion that, for example, a permanent employee attended the business premises (among other reasons).

While the collection of this information began in the second half of 2021, guidance from the privacy commissioner on this matter came only in March 2022 (OAIC, 2022). This guidance stated that evidence of an individual's vaccination status should, in most cases, only be sighted (meaning no copies should be retained and no written record should be made). Of course, this guidance came much too late.

One potential consequence of this information debacle is that health information of several thousand individuals was collected and is now sitting on a server or computer somewhere, and if that information is misused, lost, or stolen, it could have serious repercussions for these individuals.[5]

The right and responsibility of exercising personal autonomy

What this discussion reveals is how prone we are in getting caught up in our immediate environment (although this is not necessarily because of our fault). There was poor social media reporting, and the government was constantly changing orders that carried heavy penalties. The result was that we almost involuntarily created "nets" that, much like the goldfish previously mentioned, carried us away from our senses. How easily we gave in to the irresistible pull of fear; how readily we waived our right to personal autonomy.

Each individual is naturally endowed with autonomy. Autonomy comes from the Greek *auto*, meaning "self," and *nomos*, meaning "custom" or "law." Thus, to exercise autonomy means to exercise one's right to self-rule and to make decisions independent of external interference.

Shortcomings of government and civil society

In this chapter, I consider the shortcomings of the government and civil society in the context of the chaos of the pandemic. As restrictive government orders and at times poorly delivered sensational news and social media stoked our anxieties, some of us appeared to abandon our autonomy to our detriment. While it is likely that some degree of anxiety and confusion could not have been avoided even if the most efficient government decisions had been made, surely some decisions could have been made better – if not by the government, then by the recipients of the information: us.

The goldfish story revisited

Let us return to the story of the child and her goldfish. Had her parents resisted being swept up in the net of hysteria, where their mental and emotional space to move had been limited, and instead made some decisions differently, perhaps the child would have continued to live a life not too different from the one she had known pre-pandemic. At the very least, her life would not be so unbearably stressful. That was precisely the time for the parents to exercise their autonomy.

There have been a number of court proceedings brought in Australia in connection with these government orders. The court proceedings described in the next section are good examples of autonomy in action.

Challenges in court

One such case is *Kassam v. Hazzard* (2021). Various stages of the proceedings were viewed online 1.4 million times and represent the first major legal decision in Australia in relation to mandatory COVID-19 vaccination requirements. In this case, there were several plaintiffs who brought proceedings against Bradley Hazzard (the New South Wales minister for Health and Medical Research), Kerry Chant (the New South Wales chief medical officer), the State of New South Wales, and the Commonwealth of Australia.

The plaintiffs were the following: an occupation health and safety officer for a supplier to construction sites within Greater Sydney; a director of that same company; a healthcare worker employed by a pathology agency who was pregnant at the time; an employee of a hospital who worked as a private patient officer; three aged care workers employed at a retirement village and aged care facilities; a teacher employed by the NSW Department of Education; a cleaner employed by the City of Sydney Council; and an apprentice mechanic. The plaintiffs argued that the government orders constituted a violation of several rights, some of which are considered below and without reference to any particular plaintiff.

The right to bodily integrity

One was the right to bodily integrity, which the plaintiffs argued was violated when people were forced to take a vaccine they did not wish to take. The issue was that people were coerced into taking the vaccine. As a result, their consent was invalid, which meant that there was no consent at all so their right to bodily integrity had been violated. The Court decided that the orders impaired freedom of movement but not a person's autonomy over their own body, and stated that "neither provision imposed a sanction for being unvaccinated per se." The Court, however, agreed with the plaintiffs that the "orders have either an encouraging effect or even a coercive effect so far as vaccination is concerned" (*Kassam v. Hazzard*, 2021, p. 59).

Ultimately, however, the Court decided that in the context of medical treatment, consent will only be invalid when the patient has not been informed in broad terms of the nature of the procedure, where this has been misrepresented, or there has been fraud. Just because a person agrees to be vaccinated to avoid general prohibition on movement, that does not make their consent to vaccination invalid and, accordingly, their right to bodily integrity could not have been violated.

The constitutional guarantee against civil conscription

Another argument made by the plaintiffs was that the orders implemented civil conscription in violation of Section 51(xxiiiA) of the Australian Constitution. That section gives power to the federal parliament to make laws with respect to medical services, but it does not authorize any form of civil conscription. Civil conscription is the obligation of civilians to perform mandatory labor for the government. In this case, the plaintiffs argued that the government orders were a form of civil conscription because unvaccinated individuals were compelled to perform mandatory labor by providing medical services, where the "medical service" was receiving a vaccine for the purposes of contributing to the general health of the community. Ultimately, the Court decided that this argument failed for reasons that perhaps mainly lawyers would appreciate. In brief, it failed because the medical service had to be provided (e.g., by a doctor), not acquired (e.g., by an individual being vaccinated), and the orders did not compel or coerce doctors to vaccinate anyone.

Another basis on which the constitutional argument was rejected was that Section 51(xxiiiA) of the Australian Constitution only applies to the lawmaking powers of the federal parliament and not the state parliaments. Even if the vaccination mandates were a form of civil conscription, they would not be invalid due to Section 51(xxiiiA) of the Australian Constitution.

The plaintiffs' other argument was that the federal parliament *required* the state of New South Wales to mandate vaccinations. One item of evidence for this was a statement made on July 9, 2021, by the prime minister. However, ultimately, the Court decided that this was not evidence that the federal parliament required the state of New South Wales to mandate vaccinations.

This court proceeding is a good example of autonomy in action. It is irrelevant that the plaintiffs lost the case; what it demonstrates is their capacities for self-control, introspection, independence of judgment, and critical reflection. The plaintiffs were not caught by the net of hysteria swirling around them and instead utilized one of the arms of government – the judiciary – to engage in what political philosophers would call moral discourse: a discussion about what is right and wrong (Habermas, 1996). The case shows how the judicial system allows for democratic expression.

Final thoughts

On October 1, 2022, there were fewer than 1,000 reported cases of COVID-19 in Australia. Though the virus is still with us, children can now go to school and to their local park, and face masks are required to be worn in only a few places. Collectively, we are able to slow down and take a breath.

It is relevant that as the pandemic was unfolding and government orders were being made, little was known about the virus. This would have contributed to the implementation of imprecise measures based on the information available at the time. However, this essay does not presume to propose improvements to social policy decisions. Rather, it points out that we, not the government, put in place measures that were even more restrictive than the government mandates.

As a consequence, a number of hardships that befell many could have been avoided, or at least reduced, had individuals stood firm with collective common sense instead of being overwhelmed by pandemic reporting and their fears. After all, in the absence of a government order, were not parents' views valid about whether their child should be prohibited from visiting friends or going to the playground? These were decisions that we as citizens had the inviolate rights, the autonomy, and the responsibility to decide on, during the course of this pandemic. Yet it seems that at times, we failed to do so.

We will always be subject to some level of influence from sources external to ourselves (such as media reporting, government orders, and our colleagues and friends). If faced with similar challenges in the future, we should strive to act rationally rather than emotionally because, as this pandemic has shown to the detriment of our health and happiness, we will weave nets of hysteria and trap ourselves in them. It is our responsibility to recognize the net for what it is, and avoid being swept up in it by exercising our personal autonomy. If we allow external influences to affect decisions that are outside of their legitimate scopes of power, we make a mockery of our authority to determine our own actions.

We are agents of our lives

Therefore, in making decisions that are within our power, we must not forget that we are the agents of our lives. We have the responsibility for making decisions that will affect us and those around us. As parents, we make decisions affecting our children. As decision-makers in a business, we make decisions affecting our employees, volunteers, contractors, and visitors. As human beings, we make decisions that affect the members of the society we live in. This society is, after all, a *network* comprised of us. Let us together be a network of individuals who are committed to ensuring we all have the opportunity to live true human lives despite challenges we may face.

Notes

1 See Keith Storace's essay on personal development based on a growth mindset in Australia in Chapter 22, "Unlocking from lockdown: reframing the future through appreciative dialogue."
2 See Chapter 28, "Privacy issues in online education technologies in China" by Li Mengxuan.
3 See Antonio G. M. La Viña's discussion on human rights and the pandemic in Chapter 16, "Change and continuity: COVID-19 and the Philippine legal system."
4 Comparisons can be made to Li Xudong's essay on the use of technology in implementing the pandemic policies in China. See Chapter 17, "Digital technology: a best friend for implementing COVID-19 policy in China."
5 See Li Xudong's discussion on the use of technology in implementing the pandemic policies in China. See Chapter 17, "Digital technology: a best friend for implementing COVID-19 policy in China."

Reference list

Gore, J., Fray, L., Miller, D., Harris, J., & Taggart, W. (2020, December). *2020 report to the NSW department of education: Evaluating the impact of COVID-19 on NSW schools*. Teachers and Teaching Research Centre. www.newcastle.edu.au/__data/assets/pdf_file/0008/704924/Evaluating-the-impact-of-COVID-19-on-NSW-schools.pdf

Habermas, J. (1996). *Between facts and norms*. Polity Press.

Jolly, R. (2014). *Media of the people: Broadcasting community media in Australia*. Parliament of Australia. www.aph.gov.au/about_parliament/parliamentary_departments/parliamentary_library/pubs/rp/rp1314/media

Kassam v. Hazzard; Henry v. Hazzard, NSWSC 1320. (2021). www.caselaw.nsw.gov.au/decision/17c7d62628b9735ac213a597#_Toc85201771

Legal Aid NSW. (2022, January 14). *COVID-19: Breaches of the public health orders*. https://publications.legalaid.nsw.gov.au/PublicationsResourcesService/PublicationImprints/Files/991.pdf

Murphy, B. (2021). *Certification statement – COVID-19 vaccines campaign – don't be complacent*. Australian Government – Department of Health and Aged Care. www.health.gov.au/resources/publications/certification-statement-covid-19-vaccines-campaign-dont-be-complacent?language=en

Office of the Australian Information Commissioner. (2022, December 15). *Privacy guidance regarding Individual Healthcare Identifiers (IHIs) on COVID-19 digital vaccination certificates*. www.oaic.gov.au/privacy/guidance-and-advice/privacy-guidance-regarding-individual-healthcare-identifiers-ihis-on-covid-19-digital-vaccination-certificates

PART IV
Diversity in workforce behavior

19
INTRODUCTION TO THE ECONOMIC SECTOR

William Loxley and Suresh Nanwani

Introduction

Equality of diversity in the job market helps guarantee all individuals' share in the benefits of growth, including wellness in health and human services (OECD, 2021; Griffith et al., 2023). COVID-19 taught the business sector to link home and work environments to ensure workers are productive and happy while restructuring jobs to be collaborative and allow flexible working conditions (McGahey, 2021). To usher in an age of wealth and prosperity before, during, and after the pandemic, the economy is responsible for allocating resources to keep productivity strong (ILO, 2020, 2021). This includes determining trade-offs between public health and a fully functioning economy during lockdowns, given inequalities in education, finding work, business bankruptcies, loneliness, and social isolation (Beirne et al., 2021).

At the management level, businesses must rethink work structures as they vary by occupation, employer-employee relations, pay, and work-family balance (Hull, 2020; Mehta, 2021). These problems arise across labor markets that segregate the youth, elderly, women, minorities, and immigrant workers from others based on skill level and type. Skill shortages and technological advances affect the role of all workers. This is where education and training play a key role (Entwistle, 2013).

Looking at the effects of the pandemic on the business world, the practical implications of the change caused by COVID-19 have demanded reconfiguring office space for a new era, where management and supervision personnel employ a full range of human development techniques to keep employees satisfied and productive. The design of office space has to be recalibrated for social distancing and accommodating to new methods of in-and-out office work. This might imply a three-day-week arrangement or long-term out-of-office work using the latest ICT equipment. Internet access through Zoom conferencing and modern metaverse techniques will bring artificial intelligence and machine learning to the forefront of office structure as employees work remotely from anywhere.

Business travel may become less important once ICT systems move into the metaverse of three-dimensional holograms. The upper echelons of office personnel will change greatly, while work in personal services and caregiving will lead to different strategies of organization to provide quality social contact at a living wage in occupations like transport, hotels, and personal services (Christie et al., 2021).

The human element found in businesses caters to employee well-being and emotional satisfaction (Cooper, 2021). This area of human development in the business world started with work from home (WFH) when offices closed down during the pandemic. WFH allowed employees greater freedom in how they spend their work time, which often forced managers to be available up to 80 hours per week in high-pressure positions. At the same time, WFH for those without spacious accommodations caused family problems, especially when there is no available office room in the house. Parents also encountered problems of managing childcare as child services closed down, and they had to spend more time attending to children during work hours.

Similarly, career options and training were downplayed, along with issues of work compensation and promotions. This resulted in early retirements in the work-a-day world, which left a huge void to fill once office culture reopened (Hallez, 2021). Many businesses took advantage of rebuilding their workforce by adding young workers with new skill types as this helped offset labor shortages. These career issues also affected young employees starting out and left many with an appetite to go solo as an entrepreneur if possible or take a sabbatical and travel once that option became available (Nooyi, 2021).

The future of work after COVID-19 will accelerate trends in remote work, e-commerce, automation, and worker-changing jobs often. Jobs with high physical proximity are most likely to change, such as in medical clinics, gyms, travel stores, home support, and computer-based office work, especially in advanced economies (Linhart, 2022).

Six essays

Business reactions and work-related activities identified as having been impacted greatly by COVID-19 forced greater efficiency in distributing resources to modernize the work environment. Business firms and companies become interested in issues related to redesigning office space, employee-employer relations, e-commerce, pay and finance, and technological business improvements. Essay topics include work-family balance, entrepreneurial options in the gig economy, financial literacy, management flexibility, working from home issues including job satisfaction, career burnout, mental health and self-esteem issues, along with employer rights and pay (Dhore, 2020).

Reference list

Beirne, J., Morgan, P., & Sonobe, T. (Eds.). (2021). *COVID-19 impacts and policy options: An Asian perspective*. ADB Institute.

Christie, F., Antoniadou, M., Albertson, M., & Crowder, M. (Eds.). (2021). *Decent work: Better future*. Emerald Publishing Limited.

Cooper, C. L. (Ed.). (2021). *Psychological insights for understanding COVID-19 and work*. Routledge.

Dhore, A. R. (2020, April 8). *The importance of financial literacy during the COVID-19 pandemic*. www.shrm.org/resourcesandtools/hr-topics/behavioral-competencies/pages/the-importance-of-financial-literacy-during-the-covid-19-pandemic.aspx

Entwistle, H. (2013). *Education, work, and leisure*. Routledge.

Griffith, M. A., Albinson, P. A., & Perera, B. Y. (2023). Balancing rights with responsibilities: Citizens' responses to expert systems. In J. M. Ryan (Ed.), *COVID-19: Individual rights and community responsibilities*. Routledge.

Hallez, E. (2021, October 25). Early retirement surging as a result of Covid. *Investment News*. www.investmentnews.com/covid-early-retirement-rearch-213183

Hull, F. (2020). *Career after COVID-19*. Ultimate World Publishing.

International Labour Organization. (2020). *COVID-19 and the world of work: Sectoral impact, responses and recommendations*. www.ilo.org/global/topics/coronavirus/sectoral/lang – en/index.htm

International Labour Organization. (2021). *Working from home: From invisibility to decent work*. www.ilo.org/global/publications/books/forthcoming-publications/WCMS_765806/lang-en/index.htm

Linhart, D. (2022). Remote work and the contemporary workplace: The example of student internships in the context of France. In F. Rossette-Crake & E. Buckwalter (Eds.), *COVID-19, communication and culture: Beyond the global workplace*. Routledge.

McGahey, R. (2021, October 29). Is Covid-19 restructuring the job market? *Forbes*. www.forbes.com/sites/richardmcgahey/2021/10/29/is-covid-19-restructuring-the-job-market/?sh=70bbed9e48c6

Mehta, N. (2021). *Jobs and COVID-19*. Mehta Consulting.

Nooyi, I. (2021). *My life in full: Work, family, and our future*. Amazon Books. www.amazon.com/My-Life-Full-Family-Future/dp/059319179X/ref=sr_1_1?keywords=my+life+in+full&qid=1672756331&sr=8-1

Organisation for Economic Cooperation and Development. (2021). *OECD employment outlook 2021: Navigating the COVID-19 crisis and recovery*. www.oecd-ilibrary.org/employment/oecd-employment-outlook-2021_5a700c4b-en

20
ENTERING THE WORKFORCE IN THE COVID-19 ERA

S.R. Westvik

Introduction

Graduating from university is a time of fraught emotions – nostalgia and bittersweet heartache over parting with friends, excitement and determination to succeed in the working world, and nervousness at the daunting prospect of having to begin that journey. Graduating from university during a global pandemic is a decidedly different experience, marked by fear for the physical safety of oneself and one's loved ones, despair over loneliness in isolation and missed opportunities over months that can never be regained, and a deep sense of the ground shifting beneath one's feet as the global economy was upended.

As someone who graduated with my Bachelor's degree in 2020, I experienced all of these firsthand. To work through the lasting damage these years have brought on me, I sought out members of my current Master's cohort who had recently graduated to better understand the trends underpinning the generation that entered the workforce in the wake of COVID-19. I recorded the interviews in my essay, which is presented in a magazine style. The experiences of the interviewees herein represent only a very small subset of the pandemic experience for new graduates. They come from a variety of backgrounds – the United States, Italy, Venezuela, and Turkey, and myself from Singapore and Norway – but all of us have undergone the specific experience of attaining undergraduate education in European or American institutions and thereafter seeking employment or further study in these parts of the world.

It is an important caveat to acknowledge that not all 20-somethings worldwide will have had the experiences of the pandemic elaborated in this chapter. However, it is my hope that this short piece will still help illuminate the

DOI: 10.4324/9781032690278-24
This Chapter has been made available Under a CC-BY-NC-ND license.

perspectives, priorities, and challenges facing many of the young workers and students of my generation.

The last days of learning under lockdown

Studying in university, the onset of the pandemic felt like a tsunami to me – a slow drawback of water from the shore with signs visible to those who had been paying attention. Suddenly, with but a few days' warning, the wave hit with a total lockdown. I did not join the exodus of students for fear of being asymptomatic, and so was left alone in a tiny flat as the city fell silent.

March 15, 2020. I wrote the date down in the diary I kept to stay sane while emergency laws had residents locked in for 23 hours a day, excepting a designated exercise hour and quick grocery runs. That was three months before my graduation.

I should have been focused on my thesis, meeting friends for the last time, and thinking about whether I would be applying to a Master's program or applying for a job, not terrified about whether I might become one of the nearly thousand killed per day in early April by a novel disease. Watching the *status quo* unfold, I saw the future as an impassable brick wall, or perhaps a pit with no bottom that I was nevertheless obliged to jump into. The longstanding scaffold I had been prepared to climb in transition from study to work – the preparation during studies, the networking, the process of jobseeking – had been torn apart.

I was not alone in nursing this hellscape of a headspace while attempting to complete the degree I felt would be decisive for my future. Joshua King, a US student studying in his home country, felt that the social response to the pandemic negatively colored the outlook of graduates on the society we were set to enter. For him, the pandemic frightened people into having no hope, while the looting and hoarding created a sense of needing to compete to survive. The psychological impact of the coronavirus on mental well-being couldn't be underestimated.

At the time of writing – Autumn 2022, over two and a half years since the first lockdown – I am consulting a counselor for the sake of my own mental well-being, after the pandemic exacerbated old problems and, more importantly, triggered new problems such as anxiety, which I did not have to manage prior to the outbreak. These psychological impacts have a knock-on effect in the professional sphere, impacting executive function, productivity, organizational and social skills, and many other areas that should otherwise be robustly cultivated when entering the workforce.

Beyond the psychological impact of the enduring, pervasive sense of doom, many students also experienced personal sensations of academic and professional loss that impacted their transition from university to work. After all, tertiary education is a major financial and temporal investment in one's future, providing opportunities for networking (including with potential future collaborators), research, refinement of career goals, and even hitting upon tangible career opportunities, to name just a few professional benefits.

The lockdowns and halt of international movement created a profound sense of missed opportunities. For Nicola Avallone, an Italian studying in Ireland, the first year of the pandemic stripped away the joy, growth, and international connections usually fostered by going on the Erasmus exchange, arguably one of the iconic "rites of passage" for European students. He was paranoid as he wondered how the pandemic would affect his future. He decided to travel abroad for a year, which resulted in missed opportunities in carrying out research.

Joshua King also felt that the academic integrity, rigor, and prominence of his degree had been compromised due to accommodations made to assessment, including participation-style posts on forums and altered examination criteria that allowed students to pass under increased pressure. Modified assessment styles also affected Turkish student Zeynepsu Gulek, who found it difficult to retain information while working under heavy pressure to complete weekly papers. Venezuelan student Rafael Cruz Perez from the University of Central Florida felt that online classes – while necessary – deeply impeded growth, hampered the building of relationships with professors, and made it hard for students to learn and grow relationships with peers at the university. The pandemic thus significantly affected the return on investment in many students' education and careers.

Pandemic professionals: finding one's footing in one's field

COVID-19 strongly impacted students' perceptions of the potential character of their chosen career paths, as well as the work they were able to do during the "lost" years. While most people I interviewed were not swayed from their pre-pandemic plans, they raised concerns about the potential fallout of the pandemic on their careers. Joshua King wanted to go to Europe to further his studies but, as he could not go overseas, spent months at home and waited for the dust to settle.

Meanwhile, both Nicola Avallone and Rafael Cruz Perez had plans to become university professors in the future. They were unnerved by some of the consequences of shifting to online education – a clear negative trade-off, in their view, in the transition from in-office to work-from-home. Nicola Avallone noted a worrying trend that with everything online, many people had been dismissed from their positions at the university. While he was not completing his PhD at the time and therefore was less hindered by the staffing cuts, the unease surrounding online teaching and layoffs – combined with ongoing budget cuts in universities – persisted.

Rafael Cruz Perez also shared similar thoughts. He was worried as he felt his job might be replaced in the age of computers and machines. He was apprehensive that as a teacher, there might be difficulties working online with students. Though he received reassurances from family and friends that the human element will always be needed in teaching, he was not entirely convinced as he felt computers may replace teachers.

For other pandemic graduates, the shift to online work had a more positive outcome. For certain fields – such as peace and security – work opportunities tend to be centered in specific and limited metropolitan areas like London, Brussels, or New York. The complications of relocation and housing – particularly in the existing climate of inflation and a fraught housing market that significantly raises the cost of living in these major cities – can hinder the potential of applying for such roles.

There are, indeed, prevailing accessibility issues when it comes to accessing the technology for remote work, but the option of working from home helps bridge some of these gaps. For example, as a graduate in 2020, I could choose between seeking work and going on to further study. I chose to delay the latter so as to mitigate lost opportunities and was able to commence an internship with the International Crisis Group – a nonprofit focusing on early warning and analysis to prevent conflict and advance opportunities for peace. I did this without needing to worry about relocating, as I could complete it online. This role has since proved instrumental in developing my career.

Although I missed the experience of working in an office setting, I did get a chance to attain work experience in my field that may have been ordinarily trickier to do without the online option. I deeply appreciated the flexibility and the time and money I saved on relocating internationally, and now favor jobs that have even partial work-from-home options. It is true that collaborative teamwork does not have the same impact when done online – there are no water cooler conversations, for example – but I've found that I was able to develop good relationships with the tight-knit team because of proactive efforts from our supervisors and our own decisions to make intern group chats, to complete our work productively and in good company.

Zeynepsu Gulek also experienced the benefits of access that online work could provide for certain sectors. She was able to complete an academic internship, as well as commence an internship with the United Nations, both online, the latter of which continues to be a part of her life and work at the time of writing. She got used to working remotely and preferred it to working in the office. Noting the cost and legal limitations that can hinder physical relocation for job opportunities, the online job market helped some fresh graduates get a foot in the door with less hassle.

Unconventional options: making ends meet and being entrepreneurial

What about graduates who delayed their study plans but chose not to or were unable to secure field-specific internships in the intervening period? Here, work was less a source of enrichment or career progression than it was a means to make ends meet. Having delayed his plans to travel to Germany for his Master's degree, Joshua King took three jobs that were, in a way, all pandemic-related, working one of the only jobs available during the heart of the pandemic – DoorDash, a delivery service – and eventually as a substitute teacher in a local high school and with a relief branch of the US government that helped fund small businesses and storm disaster victims.

For Rafael Cruz Perez, the impact of the pandemic alongside full-time study and work left him very burned out from schooling. Thus, he elected to take a year off before going into his Master's, working as a restaurant server at Universal Studios in Florida. The experience of doing this during the pandemic was sobering: The restaurant opened up in May 2020, in the heart of the pandemic, which was a scary experience – the job had not seemed, to him, to be that of essential workers. However, many people did not seem to care about the risk.

Rafael Cruz Perez's options were limited, representing how many graduates ended up between a rock and a hard place when it came to working during the pandemic. While the job was good and much needed given the high cost of living in the United States, serving wasn't his career. Yet securing online teaching or tutoring gigs ran counter to his entire modus operandi when it came to effective teaching, meaning career-related work was at a standstill until the pandemic-related cancellations and restrictions on academic programs eased up. Thus, while some graduates were able to take early steps into their preferred field of work despite deferring some plans, others were effectively stuck in a year or two of limbo, trying to manage the economic and psychological strain of COVID-19 on their lives.

I also found that the combination of seeking online options in a transitioning work environment led me to explore opportunities in self-employment. Finding the motivation and seeking the know-how to set up a business is difficult, complicated, and stressful. However, when disillusioned with the more traditional means of making ends meet, entrepreneurship becomes increasingly an option for young individuals, particularly given the extent and ease with which business can be intertwined with social media presence. In the second year of the pandemic, I spent the summer learning as much as I could about registration and taxation for sole proprietors/freelancers in the two countries I knew I would be living in during my Master's (Germany and Ireland).

I drew on mandatory business courses I had taken in undergrad study and my experience in social media management to set up a website and begin building a small presence on some social media networks. That same year, I commenced work as an independent contractor, doing research, editing, and proofreading. There is a distinct lack of job security or stable income in doing this, of course, given that one is reliant on clients and is not salaried. Contractors typically have to pay into their social security contributions and health insurance in Europe at different rates than salaried employees, and not all countries have good incentives to support start-ups and small businesses, for example, with their tax brackets. That said, I appreciate the flexibility this has given me, allowing me to work for multiple clients in different fields and different countries, while also completing my Master's program, and to do all of this online.

I do not yet know if my small business is something I will grow or pursue in the long term. However, I find that any return to a corporate environment or office setting will have to come with more than just financial stability in order to be satisfying. Work-life balance is a major priority, as is flexibility of working conditions, and right now it is the start-ups (many of which have fully remote and highly international teams) and the companies allowing some measure of work from home that demonstrate a willingness to adapt to what keeps employees productive and comfortable.

The pandemic has given me this clear sense of perspective: Even with rising costs and even with insecurity, I would rather earn less money and do a job that leaves me feeling I have made an impact than work in a stressful environment that compromises my productivity, my well-being, and my sense of satisfaction in my work.

Entering a new world?

Now that the world is beginning to open up again, every contributor to this chapter has begun to take strong steps toward their pre-pandemic career plans. We are all currently pursuing a Master's degree in person. Some of us are doing online work on the side. Things appear to be back on track career-wise, though the lessons we learned during the past years hold steady, given the ongoing issues in 2022, where the third year of the pandemic has coincided with war in Europe, soaring inflation, housing crises in countries around the world, and of course, the ongoing climate crisis.

Writing this essay has proved to be an exercise in extending the candid conversations we already have about our backgrounds and our plans for after our studies. It has also provided a level of catharsis when reflecting on the career-related impact of the pandemic.

What have been our key takeaways from this experience? Perhaps it is understanding how fragile people are, which was a revelation for Avallone. Perhaps it was the way the stagnant months urged Gulek to read so much and research how she can improve herself online, taking her personal development into her own hands. The greatest takeaway, I believe, was best said in one of King's statements: "Take nothing for granted. . . . I never really felt [the truth of those words] until I feared for my own life at the young age of twenty-two. Not only my life; but also the lives of my family and the very existence of everything I hold dear."

We go into the professional world having now emerged from a crisis – we all lost things, but amid the struggle we gained a newfound depth of self-reflection, inner strength, and tenacity. Learning and commencing a career during COVID-19 was a deeply formative experience in learning not to take our plans and our opportunities for granted, to find a way to survive when dreams were put on hold, and to seize them with both hands when they finally came within our grasp, as they have done now.

21
HOW A GEN Y BECAME A GEN Z AT HEART

Bina Patel

Introduction

In our world today, we are experiencing many changes brought upon us since the pandemic started. This is not to say the pandemic has led to such changes, but it has been a contributing element to varying outcomes. Specifically, in our workplaces over the past two years, many were told to work from home or, frankly, not work at all – based on the nature of our professions – while others in the frontlines have worked tirelessly trying to save lives. Adjusting to change our day-to-day routines was no doubt challenging. Once the dust settled, new routines became a norm in our personal and professional lives. It has been my observation that many of us are increasingly spending time with our families and appreciating this time. What the pandemic has taught society in the United States is to value time we spend with our loved ones. In this chapter, I will share my journey and discoveries on the value of time and family during the pandemic as it relates to intergenerational differences.

My journey circa March 2020

It is June 6, 2022, as of this writing and we are in a phase where masks are no longer required. I live in Alexandria, Virginia, the United States. I have officially returned to work two days a week. The adjustment has been very hard for me. While I am probably very lucky to only need to come into the office twice a week, my mental adjustment has been rather difficult.

My journey with the pandemic began in March 2020. I was on business travel to Louisiana. While I was excited to go, I was a little apprehensive as the pandemic had many unknowns. It was in full swing by the summer of

2020, and there was a lot of chaos. I was a little anxious to travel and took every precaution necessary, including wearing the KN95 masks, washing my hands frequently, and communicating with people, when necessary, at a safe distance. I also lived in a hotel due to my job requirements. At work, everyone was tested for the virus once a week, randomly of course, to ensure we were not carrying any symptoms. In my workplace, each of us had our temperatures checked every morning. I was often in a state of surprise to see how precautious everyone was in my office.

Mind you, we were about 1,500 people in a large two-story building. After a month or so, when we had our first case of virus exposure in the office, I was nowhere near the area. This area was fully under restriction, and everyone that worked on "that side of the building" was quarantined and sent to work from their locations, mostly back to their hotel rooms. Folks on that "that side of the building" did feel as if the Scarlett letter had been placed on them. They felt scared, isolated, and stigmatized.

This was not by virtue of the leadership that was very empathetic, but a personal opinion of many. You may be wondering what my position was and what I was doing. I was a deployable alternative dispute resolution advisor, tasked to provide conflict resolution services to employees working with crises and natural disasters. In simpler terms, we worked with states that were impacted by bad weather, hurricanes, tornados, and now, the pandemic. I was on official duty working to support two major hurricanes that had hit the state of Louisiana in the midst of this pandemic.

Nearly two years post-pandemic, the world has made progress with vaccines, research, and more research to combat the variations of this initial virus. The general consensus is that people are sick of being at home, wearing a mask, and simply not returning to normalcy pre-pandemic. While many of us are "bolstered" up and vaccinated, we continue to remain cautious. You see, as of today, the number of reoccurrences is increasing in the community I live. This is because people are not wearing their masks and not taking simple precautions as before. I find that such a declaration of wearing masks may be based on a "coolness" factor, generational, and political differences.

Values and family

While the pandemic has shown me many things, the most prevalent issues are intergenerational differences. As a millennial who is on the cusp of Gen X and raised by baby boomer parents born between 1946 and 1964, I value work! Baby boomers prioritize life by placing work first, followed by family life. The "I" does not exist! With that said, each of us inherits patterns from our families. It is only once we experience life that we begin to determine which patterns we would carry forward.

A common pattern with the baby boomer generation is working long hours – over 40 hours per week. While I recall those days in my early stages of corporate life, I can safely say that I am no less selfish today than I was prior to the pandemic. In fact, the pandemic has resulted in some level of maturity. I have taken the time working from home to read, became patient and understanding with people, and prioritized myself.

Why? Great question! I switched jobs in the middle of the pandemic, a little over a year ago, and have worked remotely from home. Initially, adjusting to working from home was hard because I was physically separated from my colleagues. Later I became used to working from home and not meeting them. This was a nice feeling. Generationally, we are wired to create a routine and follow it. Millennials are born between the year 1980 and 2000. I am born in 1981.

Generation Y: commonly referred to as the millennials

According to Indeed (2022), my generation makes up the majority of the US workforce today. There are many stereotypes about my generation, particularly from Gen Xers who believe we are not responsive to feedback, lack a strong work ethic, do not take direction, and change jobs frequently. I would like to address these stereotypes. Our work ethic is strong and individualized. While it may not be a cookie-cutter way of working hard such as slaving away at our computers for over eight hours at a 9–5 job, we work with flexibility and autonomy. Hence, we have embraced remote work with open arms! While millennials take direction and feedback, we are quite vocal about stating our positions. If the feedback is unethical and incorrect, we will directly say it, and sometimes, it hurts, but we frankly don't care.

We are collaborative and, for the most part, set our personal differences aside to get the tasks completed, specifically when working in teams. While we respect authority and are very loyal to our managers, we do change jobs frequently. This is not personal. When we see what happens to the eldest employee in the room, for example, he is laid off after 25 years of hard service solely because the organization does not want to give him his retirement as it is about cost-savings, we recognize loyalty only goes so far. In fact, we also change jobs out of boredom, career expectations, and growth. We are practical thinkers and watch how others are treated, only to know we, too, will be treated the same way when the time arrives. As I mentioned earlier, I have switched jobs, and again, this is because I sought a work-life balance. As the pandemic created this space, I see it as a gift from the universe.

> *"The pandemic forced me to slow down! Instead of being on the go constantly and trying to do everything for work, I recognized the value of boundaries."*

Since my adult life began, probably my late teens, I was taught to always put my family's needs before mine. Over time this pattern is something I wished to not carry forward. I learned during these months working from home the value of boundaries. I felt guilty at first. It was hard to say no to my parents because I was working not just from home, but working a lot more!

I was working longer hours and logging in after work hours. While this was hard to disconnect, I recognized that I needed my "me time" – a break mentally and emotionally from everyone. I am an organizational ombudsman by trade. I work with people and conflict all day, every day. The days I am not doing that, I am preparing and researching for a training. I love to read, but reading fun books is also out of the question. I have to do additional research to find better and creative ways to help my workforce.

Hence, I developed a nice routine, one that also allows short 15-minute walks in between, and a longer walk at the end of the day, to de-stress. This new routine worked for me until a change occurred. The change: I am asked to come back to work. This angered me, and more than anything, it did not make sense. I work for an organization that offers 100 percent full remote work – this is an incentive – and yet I am forced to come into work two days a week. How does this make sense? It really does not! But I love what I do and so I must conform – again! More so, what truly bothers me is that I don't have the ability to decide for myself. I have to come in and that bothers me immensely because I don't agree with it.

"Thirteeners": most famously known as Gen X

While speaking with my sister who is a pharmacist, I found that her situation was the opposite. She comes from the Gen X generation, individuals born between 1960 and 1979. Gen X is a group of people who grew up during a time of social, political, and cultural change. This change resulted in Gen Xers becoming more independent and autonomous, providing them with the ability to make their own decisions. In terms of the workplace, Gen Xers make up approximately 35 percent of the US workforce.

Gen Xers work hard and complete tasks timely. They prefer independent thinking and accept responsibilities for their actions. They want to be given more responsibilities and are usually time-bound. They want to contribute to meaningful work as they value their work experience. They communicate precisely and directly. They are raised by baby boomers with a mentality of working to live. Hence, my sister, a true Gen Xer, does as she is told, including going into work.

Nevertheless, at a recent birthday party, I had an opportunity to chat with a few parents from my nieces' school who were primarily from Gen X. Several of the moms who I spoke with did not want to return to their regular jobs. While speaking to them individually, I learned they all shared a common pattern: The pandemic made them realize how much they are missing out on their children's lives. In fact, the pandemic has given them the time to reconnect with their children and spouses. More so, the mothers stated that while they had the option to work from home, almost all of them ended up resigning from their professional jobs to find careers with their hobbies: baking, working at a museum part time, making candles, and so on.

In the case of my sister, she no longer wants to be a pharmacist. Rather, she wants to open up her own bakery and café. Her passion is baking and right now she is working toward this goal. Another mom I spoke with quit her job at a well-known hospital to work part-time for another company because of their remote option, then quit altogether so she could continue being there for her kids. Working in IT, a passion once, is now a chore. She has found her new passion as a full-time mom.

Lastly, a third mom I interviewed mentioned that she opened her own part-time therapy practice for kids and is currently working remotely. She quit her full-time job as an elementary school teacher. Having her own business allowed her to set her own time while having the time to spend with her children.

What do all these moms have in common? The desire to spend time with their kids, do what they want, and just figure things out as they come. If you know anything about Gen X, it is that they are raised by parents from the baby boom generation. Their motto is to continue working, working hard, and making money "while you are young." This generation that works hard to enjoy luxury goods has prioritized family time. The Census Bureau recently reported, "In 2020, 69% of parents reported reading to young children five or more times a week compared with 65%, in 2018, and 64% in 2019" in the United States (Mayol-García, 2022, para. 14).

Even before COVID-19 started, less time was spent outdoors. During the pandemic, more time was spent inside the home at the dinner table, and thus, parent-child reading increased. The data illustrates the impact of the pandemic on parental involvement. In the case of these parents, they realized family time with their kids and spouses carried more weight. While Gen Xers prioritized their work before family, we see a shift in mentality and desires across generations to spend more time with their families.

The silent generation, labeled as Gen Z's

Gen Z's – what can I say about Gen Z's, except that they are misunderstood and I highly appreciate them. If you have watched memes about this generation, you will notice they are made fun of for wanting certain things at their own time, such as showing up to work at 10 a.m. virtually after their Starbucks run. This is a known fact. They are looked down upon by other generations and stereotyped as being "entitled." A mutually agreed-upon perception by both baby boomers and Gen Xers about Gen Z's: They walk slow, think slow, they are slow. But the largest misconception is that they are lazy. They are not lazy; they value themselves and their time. It is really about the "I," "me," and "I" alone.

They believe in inclusivity and gender fluidity, and find their online friends are the same as those they have physically. Francis and Hoefel (2018) state that Gen Z's are a touchy-feely generation that is emotive and lives on values. They do not care about the paycheck. They are the online Instagram babies, expressing their moods online. They do not believe in working eight-hour days. From my professional observation, their motto is this: Once the work is done, I can run my errands and get on with life. "Why do I need to work a full day?" Gen Z's are fast, efficient, and value personal time. They will be vocal in all that they do, despite agreeing to it or not. And if the boss has an issue, they will find another job and quit on the spot.

My cousin is a Gen Z. Yes, he walks slow, but he is a fast thinker. He does everything on his time, except when has to report to work. What I admire most about my cousin is that he has solid boundaries consistent with all, including friends and families. I have learned from him that personal boundaries are important. For example, "When eating dinner, if this is your television time, don't pick up the phone no matter who it is." They have solid boundaries – a time to meditate, exercise, exercise breathing – independent of others, including family!

My cousin, who is 26 years old, has shown me that my time is equally valuable. This means not checking emails when I am not on the clock or not picking up my phone when I eat. Or it may be something as simple as winding down one or two hours before my bedtime. Let's also mention vacation time – they need their vacation and will take it whether it is approved or not. When the flight and hotel are booked, employers better believe the Gen Z's will head out regardless of any mission-related requirements.

How have the baby boomer generation adjusted? For one, my parents appreciate the pandemic. They are scared of getting sick, both being in their mid-seventies. But my aunt, who is in her late sixties, often tells me to stop working so much and enjoy the money I am earning. My cousin says the same thing. Unlike him, however, I am afraid of breaking my lease, spending USD 4,000 to live closer to the office. I would have preferred waiting until my lease was up. The baby boomer generation as I know it appreciates family time more – they love hanging out with their kids and grandkids. They value the pandemic as something that had to occur to slow down and enjoy life. Now I hear my mother telling my sister and me to enjoy our lives! In fact, the kids in our family *want* to spend time with the "elders." Today, they value the pandemic for simply helping us to become centered.

Learning instances (also known as moments)

What I have learned from my parents and folks from their generation is that they have adjusted their way of thinking as it relates to time. With the number of deaths increasing due to the pandemic, spending time with one's family instead of working to live is more important. While baby boomers, Gen Xers, and millennials are hard workers to a degree, they are seeing an adjustment. This doesn't mean they cannot accept how Gen Z's just quit their jobs on the whim and live off of Ubering or Instacart deliveries. But with the pandemic still going on, I am seeing a new appreciation for the quality of life from the traditionalists.

In writing about the three different generations, my goal was to illustrate how much the pandemic has been a blessing. While we have experienced cries, deaths, and extremely sad moments, the pandemic has in fact slowed down society and brought people together on social issues, such as the #metoo-movement and #blacklivesmatter. The pandemic has been a revelation, and it is no longer the most important. I can safely say it has brought my family, community, and friends closer at an intimate level and created a space for my personal growth, mentally, spiritually, and emotionally. It has resulted in an open mindset at a deeper level with people and conflicts, and it has established a sense of deep patience for the follies of the world.

I can safely say that I spend more time with my family today than I have ever done in my adult life. As my parents are aging, together with my aunts and uncle, I had to find a job that has very little travel. I also want to be around to see my nieces and nephews as they grow up. With that, I find that my weekend time is often caught up with family dinners. It is nice and nostalgic. My Gen Z cousin has taught me the value of life-work, not work-life, balance with prioritizing myself and personal time. What I admire most about Gen Z's is their firm grasp to not only live a life for themselves but teach those around them, especially their parents and elders, to enjoy life before we age.

Blessings in disguise?

In Hindu culture, we believe that when we pass on, we do not take anything material with us, not even our own bodies. Only our soul moves forward. The time we spend building and strengthening our bonds is something the Western world has not adapted to, but they have had no choice. With the push to get Americans back into offices, working a hybrid work schedule, the stress to manage a previous life with the new one is not easy. In fact, work collaboration has resulted in increased levels of stress – I see this based on the day-to-day work I do when dealing with people. Yet, folks are utilizing in-person hours to catch up with colleagues or chat about their weekends, and it is more social than not. It is a nice balance, no doubt, but enough to meet the desires of building a work-life balance.

As I mentioned earlier, Gen Z's are teaching us to value life. They believe life-work balance is far more important. Slaving away at a desk all day when they've completed the work in less than eight hours is not the norm for them. In fact, this young population believes that when they finished their work in a short period of time, they should be able to take care of personal tasks. Organizations that once stated such types of people can seek work elsewhere are experiencing a situation of great resignation. Baby boomers are encouraging Gen X and millennials to enjoy life now while we are young, as opposed to when we are older.

Gen Z's are showing us this very norm, while Gen X and millennials are making the shift – quickly! I have much to learn from the Gen Z generation, including distancing myself from individuals who only tap into my energy as it benefits them. I have also created a small network of folks who have similar interests as mine. At the end of the day, I value my personal space and energy, including respecting the boundaries of others and myself. I have also become self-disciplined by not answering text messages past a certain time and not sharing my schedule with friends.

And finally . . .

In the business world, many organizations are feeling the impact of the great resignation. We are past the point of "quiet quitting," where an employee willingly does not produce above and beyond. McKinsey and Company published a report stating how organizations will need to humanize work beyond giving thank-you bonuses, flexible hours, and bump-up in pay. These incentives that once worked are no longer enough. Gen Xers, millennials, and Gen Z's desire "a renewed and revised sense of purpose in their work. They want to feel a sense of shared identity . . . they want meaningful – though not necessarily in-person – *interactions*, not just transactions" (De Smet et al., 2021, para. 2). While it is up to the organizations to determine the degree of inclusivity they will establish for their employees, they must pause to understand why their employees are running!

I have to thank the pandemic for providing me with this lesson of establishing boundaries and spending time with my loved ones and myself. And there is more to learn. As you read this piece, please take my narrative and experiences to reflect on the benefits this pandemic has had on you.

Reference list

De Smet, A., Dowling, B., Mugayar-Baldocchi, M., & Schaninger, B. (2021). "Great attrition" or "great attraction"? the choice is yours. *McKinsey Quarterly*. www.mckinsey.com/~/media/mckinsey/business%20functions/people%20and%20organizational%20performance/our%20insights/great%20attrition%20or%20great%20attraction%20the%20choice%20is%20yours/great-attrition-or-great-attraction-the-choice-is-yours-vf.pdf?shouldIndex=false

Francis, T., & Hoefel, F. (2018, November 12). *True gen: Generation Z and its implications for companies*. McKinsey & Company. www.mckinsey.com/industries/consumer-packaged-goods/our-insights/true-gen-generation-z-and-its-implications-for-companies

Indeed. (2022, September 11). *A guide to millennials' work ethics.* www.indeed.com/career-advice/career-development/millennials-work-ethic

Mayol-García, Y. (2022, January 30). *Parents and children interacted more during COVID-19.* United States Census Bureau. www.census.gov/library/stories/2022/01/parents-and-children-interacted-more-during-covid-19.html

22
UNLOCKING FROM LOCKDOWN
Reframing the future through appreciative dialogue

Keith Storace

Introduction

Although much of what we believe about ourselves is shaped over time, it is often in moments of crisis that we begin to wonder about our place in the scheme of things. Our own existential crisis may emerge as we attempt to work our way through the consequent challenges. Whether it's losing a job, a broken relationship, failing health, or something that affects the wider population, such as a natural disaster, a war, or a pandemic, they all impact four fundamental aspects of sustained good mental health: meaning, quality of life, security, and future (Connell et al., 2014). All four are inextricably linked and necessary for our emotional, psychological, and spiritual well-being. Since February 2020, these four aspects have come under threat for millions of individuals across the globe due to the COVID-19 pandemic. For others, they have become a point of reflection with an emphasis on reimagining a better future for themselves.

The longest lockdown

The accelerated spread of COVID-19 quickly became a global pandemic by the end of March 2020 (Singh et al., 2021). Lockdowns were enforced, with various restrictions and some concessions, as a way of circumventing the spread of the virus. Services deemed as essential – such as hospitals, supermarkets, and pharmacies – remained operational, although with a set of rules designed to ensure order and civility (Haug et al., 2020). Forums created to grasp the enormity of the problem and find ways of managing a seemingly out-of-control disaster became the norm. Mental health issues that surfaced were a strong focus, especially where predictions warned of an increase in anxiety

DOI: 10.4324/9781032690278-26
This Chapter has been made available Under a CC-BY-NC-ND license.

and depression, and of an impending threat to our way of life – predictions that were eventually realized (Brunier & Drysdale, 2022; Daly & Robinson, 2022).

Closer to home, Melbourne, Australia experienced six separate periods of lockdown: a combined total of almost nine months between February 2020 and August 2021, and the longest lockdown on a global scale (Wahlquist, 2021). A multicultural society with around 140 nationalities, 223 languages and dialects, and 116 religious faiths, Melbournians, for the most part, lived through the lockdowns with minimal resistance. Apart from the usual political debates that emerge in times of change, there was no obstruction to the medical, educational, and mental health support across the nation provided by federal and state governments. Despite the very welcomed assistance provided, mental health challenges continued to place a strain on mental health services due to extended periods of lockdown (Boseley & Davey, 2020).

As we now know, lockdowns, restrictions, and the eventual development and distribution of vaccines reduced and contained the spread of the virus enough to let us return to an adapted form of normal life. We still wear masks in some situations, and we still need to comply with specific rules in others, but we have reopened. Yet the impact on mental health post-lockdowns remained.

Therapy for a more meaningful future

COVID-19 has raised questions for many about life before the pandemic, and how they can reframe their future to live more meaningfully and confidently while improving their quality of life. Working as a psychologist in Melbourne, Australia, I began treating clients from the introduction of lockdowns for anxiety, depression, a perceived sense of loss of control over one's life and future, and identity crisis; all impacting well-being and quality of life (Santomauro et al., 2021). This came as no surprise.

What intrigued me though was the increase in clients experiencing anxiety who sought therapy that focused on living more meaningfully than they were pre-pandemic. Many of them described lockdown as an inadvertent blessing that provided them with time out from their lives to contemplate changes they felt strongly compelled to make. I discussed this with colleagues in Melbourne as well as Canada, the Netherlands, Brazil, the United Kingdom, and Iceland, who also reported seeing an increase in client presentations with a focus on transforming their lives. This was echoed at the European Conference on Positive Psychology (ECPP) held on June 2022 in Iceland, which had been postponed for two years due to COVID-19. I was there conducting a workshop on the appreciative dialogue (ApDi) therapy program for coaches, counselors, psychologists, and teachers, many of whom discussed their work with clients experiencing the post-lockdown phenomenon that became known as the Great Resignation.

The great resignation

Millions of people across the globe were expressing dissatisfaction with their jobs and leaving them for a more fulfilling option, with some electing to take time out altogether. Titled the Great Resignation, media reports embraced it as a positive revelation (Liu, 2022). Commenting on Melbourne's extended lockdown, Doring (2021) noted: "After almost two years of being isolated from life as we once knew it, particularly in Melbourne's case, it is only expected that people re-evaluate priorities, so why not embrace the change as a positive one?" (para. 4). This so-called Great Resignation gained momentum in Melbourne, highlighted by a drop in staff ratios across various industries. Individuals elected to leave their jobs for better options. Focusing on what matters most became more of a priority than prior to COVID-19.

Growth mindset

Many clients I worked with as a direct impact of lockdowns explained they were looking for a transformation, one where they could live more authentically. Phrases like "I feel as though who I really am is buried beneath the life I'm living" and "I feel the life I'm living doesn't truly represent who I am" were reminiscent of clients I have worked with who were struggling with severe self-doubt and low self-esteem. A clear difference though was a desire for growth in those who considered lockdown in a positive light, despite the anxiety they were experiencing. Although anxiety was impacting their motivation, they were not preoccupied with the fear of failure, as is often the case with someone suffering from self-doubt. It seemed likely these clients would have the capacity and motivation to adopt effort-oriented strategies that would help them realize a more meaning-filled way of life.[1]

Appreciative inquiry and the appreciative dialogue therapy program

Appreciative inquiry (AI) is a strengths-based and solution-focused approach to change in organizations, communities, and various types of groups (Cooperrider & Srivastva, 1999). Cooperrider states: "Appreciative Inquiry is the cooperative co-evolutionary search for the best in people, their organizations, and the world around them" (Cooperrider et al., 2008). Having worked extensively early in my career with individuals and organizations where I applied AI, I was curious to know if it could prove helpful in therapy for people experiencing limiting beliefs.

Much of my work over the years has addressed the debilitating impact of self-doubt and low self-esteem, where a person's sense of worthiness and competence are quashed by a negative view of themselves. Their decision-making capability and capacity to address challenges is overtaken by such a view, and any attempt to rise above it is quickly overwhelmed and repressed by negative self-talk. The incapacitating nature of self-doubt compelled me to delve deeper into the underlying limiting beliefs and, more specifically, into the core beliefs that were preventing people from accepting and benefiting from a more self-positive view.

Using the five core principles of AI as a starting point, I eventually developed the ApDi therapy program for higher education students suffering from severe self-doubt (Storace, 2017). It demonstrates to the client how to inquire about themselves and their situation appreciatively rather than negatively by identifying negative core beliefs, and by amplifying key strengths and values that will help them move toward what matters most to them.

I continue to use the ApDi therapy program in private practice as well as conduct consultations and workshops nationally and abroad. It has also been effective when implemented as a group program (Bell & Gill, 2018). Given its success with people experiencing self-doubt and low self-esteem, I decided to use ApDi for those who were presenting to therapy because of the impact of lockdowns. As their sense of meaning, quality of life, security, and future had been challenged during lockdown, they were prepared to do whatever to move forward.[2]

Appreciative Dialogue Framework

The ApDI framework in Figure 22.1 incorporates the five core principles of AI, a psychological structure that reinforces these principles, and four interactional stages that work between the principles and psychological structure to build toward a positive outcome. An understanding of the dynamics of the ApDi process is essential for developing a cocreative approach to therapy.

Five core principles

Every session is imbued with the five core principles of AI as they reflect the purpose of the ApDi process and contribute to strengthening the therapeutic relationship. The value of each principle is in their relevance to the kind of change necessary for a person to shift away from limiting beliefs. The principles can be summarized as follows:

The Poetic Principle: *Narratives are deeply meaningful*
The Simultaneity Principle: *Inquiry creates change*
The Anticipatory Principle: *Our image of the future guides our action today*
The Positive Principle: *Positive influence leads to sustainable change*
The Constructionist Principle: *Conversations create reality*

192 Keith Storace

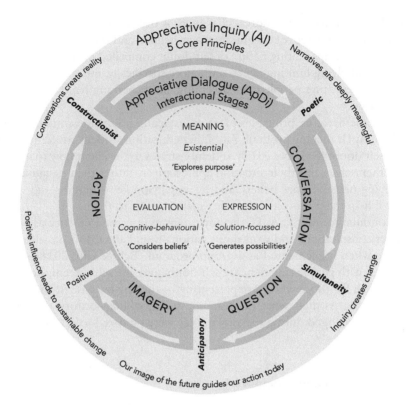

FIGURE 22.1 Appreciative Dialogue Framework.
Source: © Keith Storace

Psychological structure

The psychological structure incorporates existential, solution-focused, and cognitive-behavioral approaches. The strength of each and the symbiotic way in which they are applied throughout the ApDi therapy program encourage a level of interaction and introspection between therapist and client that supports a cocreative approach to producing solutions. Each psychological approach is labeled for its particular focus within the ApDi process and used to generate questions that will promote a shift away from what is impeding change:

MEANING – Existential | Explores purpose: *Individuals create their own meaning.*
EXPRESSION – Solution-focused | Generates possibilities: *Elements of the desired solution are already in the person's life.*

EVALUATION – Cognitive-behavioral | Considers beliefs: *Behavior change is a result of change in one's thoughts and beliefs.*

The psychological structure also incorporates seven positive stimulus statements that encourage practical and creative conversations and are examined with the client at certain intersections of therapy. Generative questions formulated during therapy, with an emphasis on the client's circumstance, are underpinned by the following statements:

BELIEF   "*Belief influences choice.*"
IMAGINATION "*We are who we imagine ourselves to be.*"
FRUSTRATION "*There is no failure, only perseverance.*"
SUCCESS "*Success is not limited to natural ability.*"
POSSIBILITY "*The positive emerges through the possible.*"
FORESIGHT "*H+I=F (Hindsight plus Insight equals Foresight)*"
ACTION "*Our goals are as achievable as the actions we take toward them.*"

Four interactional stages

Operating between, and supported by, the five core principles and psychological structure are four interactional stages designed to assist clients with stepping into a more welcomed reality. These stages are fundamental to ApDi and include:

CONVERSATION: *Engaging – Back Story/Beyond Story*
QUESTION: *Stirring Curiosity*
IMAGERY: *Exploring Possibilities*
ACTION: *Reality Formation*

It is at the Action stage where the benefit of the preceding stages coalesces to provide the client with a platform that works toward strengthening resilience and achieving their goal. Sessions move in a direction that will identify, illustrate, and maintain positive imagery that emerge and become pivotal to establishing a pathway that will lead to the desired change. Good conversations appreciate over time, and with the right kind of questions and subsequent possibilities that surface from these conversations, the client is equipped with an understanding and insight into how to move forward. Image formation leads to reality formation, and action toward realizing it provides the client with a realistic way of knowing how to build resilience and use what has been learned as a transferable life skill.

Figure 22.2 is a visual representation of the ApDi sequence that works toward unraveling and understanding what is necessary for change to take place.

FIGURE 22.2 Appreciative Dialogue Sequence.
Source: © Keith Storace

The entire ApDi process concentrates on strengthening a person's confidence and self-assurance around what can be accomplished by understanding their core beliefs, utilizing their strengths and values, challenging the notion of failure, and cocreating realistic and achievable situations. ApDi generates thought-provoking questions that stimulate a healthy outlook, conjures imagery that sets the desired direction, and supports behaviors that strengthen and maintain motivation (Storace, 2017).

Analysis and discussion

Demographic overview and assessment tools

The ApDi therapy program was initially used with 12 clients from the first and second periods of lockdown that began in March 2020. Sessions were delivered via video-link. A further 11 clients entered the program from the fifth and sixth periods of lockdowns that began in July 2021. These sessions were delivered in person.

ApDi sessions were conducted weekly, and the number of sessions varied between each client depending on their life circumstances, any preexisting mental health conditions, and goals (see Table 22.1). In summary, all 23 clients completed the ApDi therapy program with 19 attending 8 sessions, 2 attending 10 sessions, 1 attending 12 sessions, and 1 attending 18 sessions.

TABLE 22.1 Gender, age, activity status, and preexisting mental health condition

	Age			Activity status		Preexisting mental health condition
	19–29	30–40	50–60	Employed	Student	
Female	9	3	5	1	6	6	4
Male	14	4	8	2	8	3	1
Total	23	7	13	3	14	9	5

Clients who were employed either worked from home or located at their place of work if they worked in what was deemed as an essential service. Students included Undergraduate, Master's, and PhD candidates, all of whom studied online throughout the lockdown periods and often in-between these periods prior to restrictions being relaxed. Although some students worked part-time, their primary activity was study.

Prior to entering the ApDi therapy program, all clients were screened using three self-report assessments: the Beck Anxiety Inventory (BAI) (Beck et al., 1988), an adapted version of the Dweck Fixed and Growth Mindset Scale (Dweck, 2006), and the Resilience Scale (Wagnild & Young, 1993). Clients who obtained a BAI score of no higher than 25 (moderate anxiety), and a Fixed and Growth Mindset Scale score of at least 34 (growth mindset with some fixed ideas), were eligible for the program. All three scales were readministered at the end of the ApDi therapy program to provide information and discussion on the efficacy of the program for the client. The Resilience Scale was also used to explore any change in meaningfulness and motivation. Other variables that may have influenced a change in clients' anxiety, mindset, and resilience levels – such as changed personal circumstances – were taken into consideration.

A summary of anxiety, mindset, and resilience results prior to attending and following completion of the ApDi therapy program is presented below.

Anxiety

The BAI incorporates four classifications of anxiety from minimal to severe (Beck et al., 1988).

Table 22.2 highlights an overall decrease in moderate anxiety of 30 percent, an overall decrease in mild anxiety of 4 percent, and an overall shift to minimal anxiety of 38 percent.

TABLE 22.2 Comparisons of anxiety levels pre- and post-ApDi therapy based on average anxiety scores

	Score range	Number of clients Pre-ApDi	Number of clients Post-ApDi
Minimal Anxiety	0–7	1	9
Mild Anxiety	8–15	6	5
Moderate Anxiety	16–25	16	9
Severe Anxiety	26–63	–	–

Note: Clients N = 23

TABLE 22.3 Comparisons of mindset levels pre- and post-ApDi therapy based on average mindset scores

	Score range	Number of clients Pre-ApDi	Number of clients Post-ApDi
Strong Growth Mindset	45–60	4	10
Growth Mindset with some Fixed Ideas	34–44	19	13
Fixed Mindset with some Growth ideas	21–33	–	–
Strong Fixed Mindset	0–20	–	–

Note: Clients N = 23

Mindset

The adapted version of the Dweck Fixed and Growth Mindset Scale incorporates four classifications, from a strong growth to a fixed growth mindset (Dweck, 2006).

Table 22.3 highlights an overall shift from 17 percent of clients with a Strong Growth Mindset to 43 percent – an increase of 26 percent – and an overall shift of 83 percent of clients with a Growth Mindset with some Fixed Ideas to 56 percent – a decrease of 27 percent.

Resilience

The Resilience Scale is composed of 25 items in the form of statements with a seven-point Likert scale rated from "strongly disagree" to "strongly agree" and grouped into five themes (Wagnild & Young, 1993).

TABLE 22.4 Comparisons of average scores for each resilience theme pre- and post-ApDi therapy

	Score range	Average score Pre-ApDi	Average score Post-ApDi
Equanimity	3–21	9	17
Perseverance	8–56	28	44
Self-reliance	5–35	14	27
Meaningfulness	5–35	17	31
Existential Aloneness	4–28	15	23
Totals	25–175	83	142

Note: Clients N = 23

Table 22.4 highlights an overall increase in resilience of 34 percent. For each of the resilience themes, Equanimity increased by 38 percent, Perseverance by 29 percent, Self-reliance by 37 percent, Meaningfulness by 40 percent, and Existential Aloneness by 28 percent.

Discussion

The ApDi therapy program has demonstrated significant effectiveness in reducing anxiety, fostering a growth mindset, and building resilience for the 23 clients impacted by COVID-19 lockdowns. There did not appear to be any difference in the efficacy of the program between those who attended via video-link and those who attended in person. Twelve clients (five video link and seven in person) secured new jobs while six clients (four video link and two in person) modified their course of study.

The remaining five clients (three video link and two in person) decided to enjoy some extended time to contemplate what they would like to do. Clients who disclosed a preexisting mental health condition were not held back by it, with all five of them experiencing either a change in employment or a modification to their course of study.

Of particular interest is the marked increase across the five themes associated with resilience, for their relevance to clients' experience of the COVID-19 lockdowns that culminated in a strong desire to live a more meaningful life post pandemic. All five themes highlight the various aspects of resilience necessary to living more meaningfully: Equanimity is the capacity to moderate extreme responses to adversity. Perseverance is the ability to persist despite any discouragement. Self-reliance is the recognition and experience of personal strengths, limitations, and belief in one's self. Meaningfulness is the understanding that there is purpose to one's life and something to live for. Finally, Existential Aloneness is the appreciation that each person's path is unique and shared, and that some experiences must be faced alone (Wagnild & Young, 1993).

As the ApDi therapy program also works to evaluate and unpack how core beliefs impact a person's sense of competence and worth, the importance of building resilience – emotional stamina – is reinforced for the client as essential for dealing with anxiety, maintaining a growth mindset, and nurturing a healthy belief in oneself.

Another key focus of the ApDi therapy program is motivation, as it is the manifestation of what is meaningful, which gives us the "why" to what we do. The increase in meaningfulness by 40 percent and perseverance by 29 percent is especially relevant, as both relate to motivation, which was evident during and outside therapy. This increase compares to the 19 clients who sought and secured new jobs or a modification of their course of study.

The results of the three self-report assessments following completion of the ApDi therapy program reflect the outcomes experienced by clients. They commented on feeling more confident about the future post pandemic, and less anxious than they were prior to commencing the program; they embraced a strengthened belief in their capacity to successfully plan for and engage in the necessary work that will help them achieve their goals; and they valued the insight that receiving input from others, despite how overwhelming a situation may appear, is not only helpful but vital to their personal development.

Moving forward

Since COVID-19 disrupted the world as we know it, an increasing amount of people are choosing to work on unlocking a potential or desire within them that they strongly sense, and greatly hope, would make a clear difference to their lives. This is a positive sign as it not only reduces the stigma of seeking therapy, but it also opens the door for more programs such as the ApDi therapy program to explore and expand on the unique talents people can offer, especially in seemingly impossible times.

From a psychosocial perspective, when people are encouraged and supported to engage in work or study that resonates with their strengths, values, and what they enjoy, they in turn feel strong, valued, and motivated to produce good work. This ultimately benefits a person's mental health and social relationships.

As I enjoy the continued privilege of working with people in cocreating a more welcomed future, I am reminded of what my father would often say in times of difficulty: "How you live your life will make a difference to the people around you, so live a good life" (J. Storace, personal communication, 1972).

Many people continue to reevaluate their lives post pandemic, where quality of life takes precedence and living a good life is about living a meaningful one.

Notes

1 See Chapter 25, "Self-coaching for pandemic survivors" by Vikram Kapoor.
2 See Chapter 33, "A resurrection: human connections and beyond" by Suresh Nanwani.

Reference list

Beck, A. T., Epstein, N., Brown, G., & Steer, R. A. (1988). An inventory for measuring clinical anxiety: Psychometric properties. *Journal of Consulting and Clinical Psychology*, *56*(6), 893–897.

Bell, R., & Gill, A. (2018). Developing leadership confidence in Canadian students. *AI Practitioner International Journal of Appreciative Inquiry*, *20*(3), 76–82.

Boseley, M., & Davey, M. (2020, July 9). Calls to mental health services in Victoria double as strain of Covid-19 lockdown shows. *The Guardian*. www.theguardian.com/australia-news/2020/jul/09/calls-mental-health-services-victoria-double-covid-19-lockdown-strain-coronavirus

Brunier, A., & Drysdale, C. (2022, March 2). *COVID-19 pandemic triggers 25% increase in prevalence of anxiety and depression worldwide: Wake-up call to all countries to step up mental health services and support*. World Health Organization. www.who.int/news/item/02-03-2022-covid-19-pandemic-triggers-25-increase-in-prevalence-of-anxiety-and-depression-worldwide

Connell, J., O'Cathain, A., & Brazier, J. (2014, November). Measuring quality of life in mental health: Are we asking the right questions? *Social Science and Medicine*, *120*, 12–20.

Cooperrider, D. L., & Srivastva, S. (1999). *Appreciative management and leadership: The power of positive thought and action in organization*. Crown Custom Publishing.

Cooperrider, D. L., Whitney, D., & Stavros, J. M. (2008). *Appreciative inquiry for leaders of change* (2nd ed.). Berrett-Koehler Publishers.

Daly, M., & Robinson, E. (2022). Depression and anxiety during COVID-19. *The Lancet*, *399*(10324), 518.

Doring, B. (2021, November 26). Why a return to "normal" is sparking career changes down a more rewarding path. *The Sydney Morning Herald*. www.smh.com.au/business/workplace/why-a-return-to-normal-is-sparking-career-changes-down-a-more-rewarding-path-20211125-p59c3c.html

Dweck, C. S. (2006). *Mindset: The new psychology of success*. Random House Inc.

Haug, N., Geyrhofer, L., Londei, A., Dervic, E., Desvars-Larirve, A., Lorreto, V., Pinior, B., Thurner, S., & Klimek, P. (2020, November 16). Ranking the effectiveness of worldwide COVID-19 government interventions. *Nature Human Behaviour*. www.nature.com/articles/s41562-020-01009-0

Liu, A. (2022, July 4). What is the great resignation? *Psychology Today*. www.psychologytoday.com/au/blog/therapists-education/202207/what-is-the-great-resignation

Santomauro, D. F., Herrera, A. M., Shadid, J., Zheng, P., Ashbaugh, C., Pigott, M. . . . Ferrari, A. (2021). Global prevalence and burden of depressive and anxiety disorders in 204 countries and territories in 2020 due to the COVID-19 pandemic. *The Lancet, 398*(10312), 1700–1712.

Singh, S., McNab, C., Olson, R., Bristol, N., Nolan, C., Bergstrøm, E., Bartos, M., Mabuchi, S., Panjabi, R., Karan, A., Abdalla, S. M., Bonk, M., Jamieson, M., Werner, G. K., Nordström, A., Legido-Quigley, H., & Phelan, A. (2021). How an outbreak became a pandemic: A chronological analysis of crucial junctures and international obligations in the early months of the COVID-19 pandemic. *The Lancet, 398*(10316), 2109–2124.

Storace, K. (2017). Appreciative dialogue: Managing self-doubt through inspirational discourse. *AI Practitioner International Journal of Appreciative Inquiry, 19*(3), 65–75.

Wagnild, G. M., & Young, H. M. (1993). Development and psychometric evaluation of the resilience scale. *Journal of Nursing Measurement, 1*(2), 165–178.

Wahlquist, C. (2021, October 2). How Melbourne's "short, sharp" covid lockdowns became the longest in the world. *The Guardian.* www.theguardian.com/australia-news/2021/oct/02/how-melbournes-short-sharp-covid-lockdowns-became-the-longest-in-the-world

23
COVID-19 AND MOVING TO THE NEW NORMAL

Victoria Márquez-Mees

Introduction

Two and a half years have passed since the COVID-19 pandemic sent almost the whole planet home. Although in my case and that of many, the threats and restrictions suffered are now only a bad memory, life today is far from normal.

Our life has changed, and we are yet to see if the new normal is better than the old normal, and whether our survival kit is fit for purpose.

I have gone through these two years, first denying any potential impacts of COVID-19 in my life and work, then reconsidering it all, and accepting that I am not immune to the dramatic change in our everyday environment.

Let me acknowledge that I am an extremely fortunate individual and I am sharing my experience just to add a personal perspective. I am not expecting anyone to commiserate with me. I have not lost anyone to COVID-19, I still have a job, my family is well and thriving, and I am bound to be a more resilient individual thanks to this unexpected event.

Super-stressors

It is common knowledge that some events in life, known as super-stressors, can make you spin and lose all the carefully constructed balance in your daily routine. In February 2020, my husband and I decided to voluntarily go through two major super-stressor events – change jobs and move to a new country – little knowing that this would involve a greater upheaval than we had bargained for.

The decision to change was not taken lightly. As a seasoned economist, I know that, in business and in life, the best decisions are the ones based on full and robust information. If information is limited, your actions might not deliver what you expected. Often, if the context is uncertain, there is a preference to wait until you can better assess the consequences.

In general, only those that thrive on risk and fools are willing to act during times of uncertainty. It turns out I am a bit of both. I am a risk taker. I have changed jobs many times and moved to many different countries during my lifetime, with excellent results. I am also a fool and have suffered for my foolishness several times, as I get caught up in the allure of new challenges and miss the huge red flags.

In this occasion, both the fool and the risk taker took over. In the early months of 2020, just before the March lockdown, I was sure I had a robust set of data to support our decision to move. I knew from experience that the job change would carry some adverse impacts but would be temporary and manageable, as they had been in the past. My new employer was a multilateral development bank, the same as my former employer. My role was almost identical to the one I had been performing for almost a decade, with the added safeguard of the freedom to build the office from scratch based on my knowledge.

Furthermore, as a woman in her late fifties, I thought that this was the best moment for a change. Waiting would only reduce job opportunities and limit the possibility of advancing my career a bit before my retirement.

In all this process, I had not computed the impacts of a global pandemic of a magnitude never experienced by this generation. As many did at the time, I dismissed COVID-19 as a short temporary event that would surely be over by July of that year. I could not have been more wrong.

COVID-19 transformed life and work

COVID-19 was not a short, localized shock. It transformed our way of life and work immediately, sending us into lockdown and reducing our contact with other human beings to a minimum. Then, as those initial milestones (e.g., the summer saw a reduction in restrictions and governments were keen to reopen businesses) were reached and the risk persisted, we had to rethink our whole approach to living, bestowing upon us long-lasting changes on how we operate, the way we engage with each other professionally and socially.

Due to COVID-19, many people lost loved ones, others had their health compromised permanently, and most felt the need to change their daily routines to maintain their sanity. From my perspective, the reaction has been to set individual welfare at the forefront of our life goals in an instinctive survival fashion. However, by doing so, we have not dedicated time to ponder how some of those hastily adopted changes will affect society and us in general.

COVID-19, and the changes in the *modus operandi* in work and everyday life, have made it more difficult for me to adapt to my new environment and thrive. Among the many small things that are different now, there are three that stand out as huge hurdles to adaptation and wellness.

Lack of a social network

A social network is defined as a group of friends, acquaintances, and coworkers connected by interpersonal relationships (Merriam-Webster, n.d.). When you move to a new country, you give up your social network. It is not that you lose your friends and family. They are still there in your old neighborhood living their lives. Your old workmates still engage with each other as they have always done. You are the only one missing.

We left that network by choice. However, our intention for moving was not to isolate ourselves from others. It involved the expectation of creating a new network. As Maria Cohut (2018) said, "Enjoying close social ties – with friends, partners, or family members – can make us happy and improve our overall life satisfaction" (para. 32).

Because I had moved many times before, I knew it usually takes a year to build a new network of friends that allows the new place to be called home. However, in this occasion, all our past knowledge proved useless. Moving in Covid times was like nothing we had experienced before. Our arrival to the United Kingdom was completely atypical. Lockdowns, social distancing, and working from home prevented us from having any formal or informal interactions with others. Socially, COVID-19 has bestowed a permanent fear upon people of getting sick. Some acquaintances still hesitate when you invite them over; others cancel because they have caught COVID-19. A permanent rain check accompanies every plan.

My partner and I's first "date" with locals took place in a park six months after our arrival. Even now, almost two years after our landing, we still feel that home is somewhere else, not here. The feeling of loneliness is now our new normal, and it has made us question if things will ever return to what we knew.

Working from home

My new job meant a relocation to London. However, when I started working for my new employer in July 2020, I was still living on the other side of the Atlantic Ocean as travel restrictions were in full swing and there was no processing of visas. My days would start at 3 a.m. to fit UK time. I would sit in front of my laptop and move from one Zoom meeting to the next, trying to immerse myself in a new institution and learn from those short and formal conversations. It was not until September 2021 that I finally stepped into the office. By then, I was no longer a new joiner but still I knew almost nothing of the culture and the people. As of September 2022, we have not yet returned to the office in full. I go to the office two days a week, and the other three days are spent in my garden shed. In my case, hybrid work has brought little joy. I feel unable to establish a routine, and everything seems to take longer than before – conversations with colleagues lose continuity and, in general, I feel I am not at my best.

All around the world, many people have embraced working from home as the opportunity to balance life and work. I am not part of that group. I fear the isolation of working from home and miss the learnings, jokes, and experiences I used to get in the workplace. My situation has very little to offer, only my own ideas. Work is a huge part of my life and has been instrumental to my growth as a person. I am relieved when I see that I am not the only one who thinks this works. In an article by *The Economist*, Jamie Dimon, chief executive of JPMorgan Chase, is quoted saying: "Working from home doesn't work for people who want to hustle, doesn't work for culture, and doesn't work for idea generation" (Bartleby, 2021, para. 5). I agree with him.

As I constantly hear colleagues saying they are more productive working from home, I would like to ask for the metrics used to sustain this claim. While we might produce more reports and respond to more emails, I am not certain all that frenzied isolated activity has increased productivity in all lines of work.

Leading a small team, I find that each one of them has a clear idea of what they want to achieve. However, they have limited understanding of what others do or how collaboration might bring a better outcome. I lead an office that seeks to address the concerns of project-affected communities. For these communities, one more email or a tidier report means nothing. What they seek are for diligent people to remedy the harm. This requires quitting the home, going to the office, traveling to project sites, and engaging one-on-one.

When we work from home, we forget that many others face a different reality. We make assumptions based on what is happening to us and usually fail to understand the reality on the ground. Our individual perceptions take over when our work requires being empathic and being a good listener to identify the real issues behind a complaint, so we could provide effective remedies.

COVID-19 has made us change our way of working, but more importantly, it has changed the way we wish to interact with others. We are embracing the "I," as we do with selfies in social media. Since our focus is on individual production, we tend to disregard the value of collaboration and teamwork. The new normal is transforming the job market. We are now emphasizing individual delivery, independent work.

I go further and consider that the new normal may be pointing us toward a transformation of employment from full-time employees to a contracting of deliverables. If we value more the outputs that an individual can produce, why would we need to work full time? If I can hire any capable individual and establish fixed terms of reference and specific outputs for his or her work within a determined timeframe, is there any reason for engaging with them beyond the boundaries of a product-specific contract? Do those that resist returning to the office see that their willingness to embrace a new way of working may also include a change in contracting rules? A model where you are only paid for what you produce, where you are not required to be full time, but also do not merit benefits?

Team building

The third challenge was something I embraced with glee in early 2020 – when I accepted the job offer – but which Covid turned into a Himalayan climb. I was charged with establishing the project grievance mechanism of a multilateral development bank per a policy recently approved by the bank.

My work during the first few months was to define an organizational structure, draft job descriptions, and recruit a small multidisciplinary and multicultural team. Interviews took place via Zoom, and those selected joined the team virtually. We were the foundational team of a new office, seeking to establish a new procedure in an institution that none of us knew. The first year, my job appeared to be a game of the blind leading the blind. Some of them eventually gave up and quit before we got to meet in person. New employees craved for certainty in a world where everything was uncertain.

Informal conversations were impossible, and discussing foundational issues through Zoom was a protracted and boring process. We did not know each other, and to be a team there was a need to establish some sort of relationship as a first step. Due to my culture and origins, I thrive in an in-person environment. I am more relaxed and willing to spend time talking over a problem and finding common solutions. My virtual persona is completely the opposite. I am harsh, my emails are short and to the point, and transferring skills through Zoom or Webex is something I still have to learn to do. This made it more difficult for my team to engage and understand how important their contributions were to me and to those who seek our help.

Building a new office and a new team was already challenging in normal circumstances, but the lack of in-person contacts and informal everyday engagement made it even more so. Particularly, we were unable to quickly build the trust that comes from daily camaraderie, which is so much needed when you are creating something new. Although we are now slowly coming to that point, one of the side effects of COVID-19 is that everything has taken more time than we initially thought we needed.

When I decided to move to London, I was sure that I was up to the challenge. I knew my subject, had experience building organizations, and thrived in new environments. I had no doubt that with dedication and commitment, this new enterprise would be as successful as others had been in the past.

I certainly did not hit the ground running in July 2020. Most of my time was spent planning at a time when plans changed every day, and trying to provide certainty in an environment where uncertainty was the norm and there appeared to be no clear answers to many questions.

COVID-19 has left us with a sense of lost time and limited progress. A two-year void, which we are now frantically trying to fill. Despite all, we managed to set up the office, but it is only in recent months that we started to work as a team. Talk about long Covid.

What is next in this new normal?

My experience as a manager during Covid times has left me wondering: Is the current workforce prepared to perform in this new normal? In labor economics, the concept of structural unemployment relates to a mismatch of people's skills with those required by the economy due to structural, economic, or technological changes. COVID-19 has impressed a structural-technological change in the way we work. We have embraced the new normal under the assumption that all employees can transition from the in-person working model to a remote or hybrid model.

We are asked to work differently, but we have not had the time to develop the skills to do so. We apply our old skills and are surprised when things do not work out as expected. I look at my skills toolkit and know that some of my tools need to be disposed of, but I am still trying to identify the new skills that would replace them. Are we moving toward new rules of employment where full-time staff vanishes, and a reduced managing team is responsible for a large cohort of contractors? Then, is that same team going to be responsible for innovation, for transforming the business mission into a profitable results-oriented reality?

Are we sure this is the path we want to follow? I venture to say that no one really knows. The new normal finds us still struggling with old habits and new expectations. Job adverts must now specify if the person is expected to come to the office or be completely remote. HR departments are trying to find a sweet point on work-life balance, remuneration, and taxation issues. Managers are dedicating more time to juggling the team schedule to fit the expectations of each of the member than the business itself.

As a seasoned professional and manager, I advise caution. The decision to transform our work methodology after a two-year shock is not a good idea. Good decisions need good, solid information. We do not have that right now. Furthermore, work during the COVID-19 pandemic was not normal. Why are we then considering that this is the path to follow? COVID-19 has transformed the world, and despite many slogans, we are not building better. We, as inhabitants of this planet suffering from shock, are more fearful, more hesitant to act, more individualistic, and less collegial.

But that is not who I am, nor who anyone is in reality. We need some time to recover and then to give some thought on what the learnings of a short intense shock like COVID-19 have brought to our work practices. Perhaps, as suggested by psychologist Susan Pinker (2020), what we need is to include in our vaccination program direct person-to-person contact to release the cocktail of neurotransmitters that reduce stress and anxiety. For me, at least, that would be the best antidote to COVID-19 and the perfect context to start strategizing and innovating.

Reference list

Bartleby. (2021, November 6). Why executives like the office. *The Economist*. www.economist.com/business/2021/11/06/why-executives-like-the-office

Cohut, M. (2018, February 23). What are the health benefits of being social? *Medical News Today*. www.medicalnewstoday.com/articles/321019

Merriam-Webster. (n.d.). Social network. In *Merriam-Webster.com dictionary*. Retrieved December 24, 2022, from www.merriam-webster.com/dictionary/socialnetwork

Pinker, S. (2020, April 24). *What makes social connection so vital to our well-being?* [Audio]. Ted Radio Hour. www.npr.org/transcripts/842604367

24
FINANCIAL LITERACY

Its relevance in the education curriculum

Marie-Louise Fehun Aren

Introduction

Like the legendary dust devil in African folklore who is a demon causing harm, COVID-19 took the world by surprise. Its lingering effects still affect individuals, communities, nations, and the world at large. On the one hand, COVID-19 was a catalyst that led to some positive changes. For example, it amplified citizens' demand for public accountability and increased opportunities for remote learning through communication technology platforms.

On the other hand, COVID-19 triggered and exposed many hidden systemic vulnerabilities in nearly all areas of human enterprise, from health to education. One of the most profound changes the COVID-19 crisis brought about was a general paradigm shift – from old ways of perceiving, engaging, and responding to circumstances, to novel ways of perceiving, engaging, and responding to circumstances at individual communal and global levels.

While these shifts are still occurring in different areas of the economy, a critical vulnerability to human survival and development has been laid bare by the COVID-19 crisis. This factor refers to the socioeconomic menace of mass financial illiteracy – a near absence of prudence in managing money and financial resources, exacerbated by a glaring absence from the formal educational curriculum.

It is true that national and social discourse on urgent matters like health, vaccines, rising debt levels, economic growth, poverty, unemployment, and human rights are relevant. Conversely, it is self-defeating to ignore financial literacy as one of the building blocks of development and sustainable poverty eradication. Consequently, inclusive education should foster personal understanding of how money is created, works, and grows. In the subsequent sections, I will discuss the relevance of financial literacy taught in schools.

Financial literacy and its connection to the educational curriculum

The high level of financial illiteracy seen in many societies, especially in several African societies like Nigeria, appears to be rooted in the faulty philosophical foundations underpinning the basic educational system. According to a financial literacy baseline survey report done in 2015, Nigeria's financial illiteracy level figures are 31.1 percent in urban areas, 68.5 percent in rural areas, 48.5 percent of men in urban areas, and 50.3 percent of men in rural areas (CBN, 2015). The system appears to be designed to prepare people for employment and self-employment without corresponding education on managing income earned.

As a result, the educational system primes its students to acquire knowledge and skills that ensure their transformation into model employees, small-scale producers, and labor supplies for the overall prosperity of the few wealthy (the Pareto Principle of 20 percent who hold 80 percent of the national wealth). The economic system appears to ignore its adherents to personal financial well-being, relegating it to the realms of sole personal responsibility.

In Nigeria, the previous 1984 educational curriculum was based on the philosophical trappings of a free and democratic society, and a buoyant and dynamic economy among other values. Indeed, the 1984 curriculum worked because educational skills provided an opportunity for social mobility from a lower to a middle-income status. Yet despite these positive outcomes, the economy could not accommodate everyone in employment, leading to high unemployment levels. In the 2000s, the basic educational curriculum was revised into the National Economic Empowerment and Development Strategy (NEEDS), in response to the technological advancement of society and international sustainable development goals.

The theoretical background of the NEEDS system includes value reorientation, poverty eradication, job creation, wealth generation, and citizenry empowerment. The NEEDS system functions on the Lower, Middle Upper, and Senior Basic Education Curriculum (BEC) grades (NERDC, 2008). The projected outcome of the NEEDS system is self-reliance. Despite the obvious synergy between NEEDS and BEC financial management relationship, NEEDS is not included in the general curriculum nor taught at the senior BEC level.

The issue with relegating financial literacy to the realms of personal responsibility is due to the students' lack of time to invest in financial literacy development, since immediately after formal education completion graduates are busy seeking jobs in the formal sector. The educational system, together with the labor system, informs the general attitude about making money, spending, budgeting, work purpose, and so on. Without the required skill to administer finances and make use of opportunities provided by financial activities to increase financial well-being, many people are unable to earn a living outside of regular employment. The pandemic clearly revealed the sorry state of many adults living on the financial edge from paycheck to paycheck, living on the precipice of poverty. On the brighter side, the opportunity to make financial literacy widely available as a basic-level educational requirement is a positive outcome of the pandemic.

Factors heightening wide-scale financial literacy during COVID-19

Certain factors contributed to better preparing the public's financial literacy skills:

The irrelevance of optional financial management skills

Prior to COVID-19, times seemed relatively simple and the entrenched system of work and finance appeared manageable. Many people acquired some form of formal education to enlighten and equip themselves for the job market, with no lessons on how to prudently manage income received. Although many educational institutions did provide lessons on basic commercial courses – including mathematics, business studies, economics, and commerce – personal finance topics such budgeting, investing, cash flow, and credit management were not covered in these lessons.

Finance, investment, and wealth building were regarded as the exclusive preserve of the wealthy elites or at worse, boring. In addition, lessons on personal finance were expected to be taught at home, forgetting that lessons taught on finance are highly dependent on the quality of home-life and relevance attached to financial knowledge in various households. The COVID-19 crisis revealed that teaching individuals, especially African children and young adults, about personal finance management and skills is essential to their future financial success and ability to survive pandemics. Financial literacy equips them with the skills to make prudent financial decisions, plan their financial future, and prepare them for managing periods of financial surplus and scarcity. COVID-19 triggered a paradigm shift from the importance of acquiring financial management skills to raising its national and global prominence.

Poverty and consumerism

During the worst of the pandemic, it became clear how financial illiteracy and consumerism prior to the pandemic contributed needlessly to financial suffering. Many adults lost their jobs, and some were unable to get replacement work, feed their families, or access social grants, thereby increasing poverty. COVID-19's widening effect on poverty and inequality inspired several educators, businesses, government agencies, and social groups in some societies to become more concerned with financial literacy in the education curriculum. Studies conducted on the financial literacy of general consumers showed that financial literacy is abysmally low among consumers (Herman & Bala, 2007).

The reason for greater concern for financial literacy in schools is not hard to fathom. The financial health of individuals influences the general well-being of the society. It influences health, family stability, crime prevention, and justice reform. Poor financial management can lead to more anxiety and stress in life.

Increased personal and household debt levels

There appeared to be an increase in rising personal debt, especially credit card debt, from both COVID-19 and pre-Covid debt obligations that stalled repayment. The more personal debt acquired by individuals, the more anxiety and health problems a person is likely to have (Drentea, 2000). Prior to the pandemic, debt was always a part of many adult finances even when there was a temptation to borrow for growing financial obligations like education costs or healthcare. The unproductive personal debt buildup may have been the result of aggressive policy to make consumer credit more accessible, especially through short-term loan for items like cars, washing machines, and so on, and the huge interest rate and surcharges it commands.

Many of these loans are not income-producing loans that can justify the high interest rates. Undoubtedly, consumer credit is useful for their accessibility and convenience value. However, because neither the household nor the educational system teaches financial literacy about income-producing debt and bad debt that feeds an unhealthy spending while producing no income, many people are unable to exhibit financial responsibility when borrowing.

This is all the more potent when aggressive debt is made available to the financially uneducated. Repayment becomes difficult amid rising expenses or rising unemployment, which blocks additional income when repayments falter. During COVID-19, even though the overall level of commercial activities reduced from the lockdown restriction, the cost of healthcare, especially in systems where the public health care is unreliable, coupled with increased unemployment contributed to increased personal and household debt. In some cases, spending on food was reduced as debt obligation, demanding the diversion of funds from essentials to settling obligations.

Household debt as a percent of nominal GDP was around 41 percent in April 2020, peaked at about 45 percent at the beginning of 2021, and returned to 41 percent in January 2022. Lessons prior to the pandemic on financial skill and personal debt management may have been helpful in managing pandemic-related personal debt. Increasing inflation especially on essentials contributed to the debt problem. This is because as prices of essentials continued to rise, many people were forced to borrow money to meet the rising cost of essentials, with bleak hope of due date repayment from rising unemployment (Lusardi & Mitchell, 2014).

Financial losses from investment fraud

During the COVID-19 crisis, reports show that there had been a dramatic increase in get-rich scams and other forms of investment and employment fraud. Scammers took advantage of people's distress, leading to more personal financial loss for those lacking financial skills or knowledge. These problems helped solidify the importance of financial literacy teaching requirement in education among the public. For example, in May 2020, high anxiety occurred because of the intensity of lockdown when the public thought COVID-19 would last forever. There were lots of tempting online investment offers with mouthwatering returns, especially in cyberspace with blockchain investments such as cryptocurrencies, that came like quick fixes to the uncertainties brought on by the pandemic.

Investment scam reports surged by almost 32 percent during 2020, with losses to these scams increasing 42 percent to GBP 135.1 million in the United Kingdom. In Nigeria, many fraudulent COVID-19 schemes were through loans, relief packages, and internet data scams. A popular one was the clickbait "free 16GB data," were clicking the link would grant hacker access to the banking details of the victim. While some people fell for these scams, the financially savvy did not because their saving grace was a sound financial and investment knowledge that helped with analyzing the investment claims. These life-skills are certainly not learned from basic formal education system because the curriculum does not include finance studies.

COVID-19 and the shifting personal financial literacy paradigm

COVID-19 and its disruptions caused many people to reevaluate their personal priorities including finance. It seems personal finance knowledge goes beyond just knowing how to earn a higher income. Personal financial literacy includes the capacity to show discipline and prudence in money management that creates sound financial decisions. Consequently, financial discipline involves the attitudinal shift from false and toxic optimism on the uninterrupted availability of financial resources, to a cautious approach to managing financial resources as well as financial preparation and budgeting, monitoring of income, keeping records, managing personal debt uses and levels, and more (Aliche, 2021).

Essentially, it is living below one's means and investing resources, and investing whatever is left in passive investment income that might generate more money. For instance, COVID-19 solidified my habit of paying myself first – about 15 percent of my income, 5–7 percent for emergency funds – then budgeting the remainder for different type of essential expenses and very limited luxury items until my income stream increased.

Prior to COVID-19, I loosely observed the financial literacy of many people, especially the youth in Nigeria. This observation is in line with the Central Bank of Nigeria's reflection that the low level of financial literacy acts as a major barrier against the public's ability to manage their financial resources effectively (Central Bank of Nigeria, 2015). The overall attitude seems to be a nonchalance to systematically learning how to grow wealth. While many people have the desire to have a huge income, their desire to develop financial discipline (with regard to spending on unnecessary consumables and investing excess income) is generally absent.

The lack of interest in personal finance by students could have been because of the erroneous thinking that earning high salaries is enough, or that understanding personal finance management is boring or is the exclusive preserve of the rich. Many people also think the business of finance should be left with the financial or investment manager.

Since COVID-19 arrived and largely receded, there has been a surge in interests to take personal finance management seriously. This is attributed to private organization literacy programs. For example, organizations like QNET, a global e-commerce and direct selling leader, unveiled a financial literacy capacity building program in Nigeria, targeting 6,000 young people. Others, like the African Centre for Citizens Orientation (ACCO) with Women's Financial Literacy and Empowerment Network, launched an All Citizens Financial Literacy Program that aims to teach disadvantaged young men and women optimum utilization of available financial instruments. Other people enroll in finance classes on channels like *The Financial Diet* on YouTube.

The world of remote work created a trend of earning more income beyond the regular 9–5 salary. Social media and content creators are earning good money from their creative efforts. Although global demand has risen, people are still prudent about their spending choices. Many people believe that COVID-19 has been an eye opener. In discussions, others mentioned that the COVID-19 pandemic has taught them a painful lesson in personal finance management. Based on this feedback, it is time for the education curriculum to evolve and translate the public need to personal finance management knowledge into a financial literacy program. Understanding finance should be both a personal responsibility and the public responsibility of a nation or state through its formal channels of learning and training. Failure to respond to the call for increased public financial literacy worsens existing socioeconomic inequalities and jeopardies the capacity to manage the financial impact of future crises – whatever they might bring.

Financial ignorance carries enormous personal and national costs. The absence of understanding basic financial concepts cripples the ability of the public to make decisions related to financial management. This in turn determines poor financial choices made in the key areas of personal finance, including saving, investing, borrowing, which leads to greater personal debt and lesser personal savings.

Personal finance components

There are five core areas of personal finance: income, budgeting/spending, savings, protection of income, and investing (Turner, 2022). Figure 24.1 illustrates the personal finance components.

Income is the substance of personal finances because it is the source of a person's cash flow. It includes salaries, monetary gifts, dividends and investment income, and pension. *Esusu* is the Nigerian word for savings club, where money is contributed by members into a pool and collectively given to members until all receive from the pooled contributions. *Spending* includes funds from income set aside for expenses. Expenses therefore satisfy needs. This is one activity prior to COVID-19 that many people failed to exercise discipline in. No matter the amount of income one earned, if he does not control

FIGURE 24.1 Personal finance components.

Source: © Marie-Louise Fehun Aren

his spending, financial planning would be futile. Therefore, controlling the amount of money spent allows a person to set aside money to plan one's financial future.

Savings is purposely setting money aside for future use, either for investing or for spending. Saving is helpful for meeting emergencies. During the thick of the COVID-19 pandemic in 2020 and 2021, some of the personal financial hardships many people experienced arose from meager savings. When work layoffs increased from the national lockdown restrictions, especially in the tourism and entertainment sectors, it became difficult for many people to meet financial obligations.

Investment uses savings or earned income to create additional income streams through good rates of returns (RoRs) from purchases of real estate, shares, business enterprises or franchises, and so on. One of the values of investment as a part of personal finance skills is that it hedges the time value of money that is eroded by high inflation. The basic lessons on finance and trading I received from my father and my independent finance learning from books I read as an undergraduate allowed me to invest in commercial bank shares during the Nigerian banks recapitalization and consolidation mergers of 2005. Over the years, this income grew, and dividends from these investments helped augment my limited income 14 years later, during the rough period of the COVID-19 crisis. Investments, however, come with a risk of low returns.

Investment risks bring in another element of personal finance – *income protection*. Risk protection includes all measures taken to ensure security of investment and income. These may include buying safe investment products like government bonds, life insurance, hedge funds, and so on. The riskier the investment, the higher the returns, and vice versa. However, the form of investment protection depends on an individual's risk appetite. For example, in managing income, some people implemented a strict spending regime on essentials during the pandemic.

This helped them gain extra money to invest in financial products as a hedge for any future financial crisis. Thankfully, as the extreme danger posed by COVID-19 abated due to massive global vaccination from 2021 onward, there has been some relaxation of strict spending regime by the financially savvy. For instance, at the tail end of 2021, my spending habit extended to include a few luxury items. Financial discipline should be balanced with enjoying the good and beautiful things that life has to offer within one's reach.

What can be done?

The question becomes what can be done at the national, communal, and individual levels to include personal finance into learning systems. At the national educational level, the first step would be to invest in training or employing knowledgeable educators to teach economics and personal finance from the primary school level up to tertiary level. For example, South Africa has the South Africa National Consumer Financial Education Strategy 2013 that helps financial consumers improve their understanding of financial products, models, and risks. At the regional level, the OECD/INFE international network of financial education guidelines for Financial Education in Schools 2012 exists and provides high-level non-binding international guidance, to assist policymakers and relevant participants design, introduce, and apply useful financial education programs in schools.

A second step should be to integrate mandatory macro/micro-economic courses on personal finance management and cash flow into the core curriculum and then assess progress made. A good example is found in the US National Standards for Financial Literacy, which provides a framework for teaching personal finance in kindergarten through 12th grade. Another example is a simple curriculum created by the Federal Reserve Bank of St. Louis called "A 'Standard' Personal Finance Curriculum." Together, these curricula include subjects on earning income, saving, credit, and insurance.

In the same vein, the basic education system should include lessons that focus on earning and income management, budgeting and spending, cash flow management, spending financial analysis, credit and debt management, savings and investment decision-making, retirement planning, and so on (Herman & Bala, 2007). At the communal level, the government may provide funding to support state and local government efforts to work with development agencies and civil societies to promote personal finance education to disseminate valuable information to the public on personal finance literacy. This includes the use of radio, television programs, and social media platforms to sensitize the public about the relevance of financial literacy skills.

At the personal level, it is important for everyone to learn about finance to augment formal teaching methods. This can be done through reading books on personal finance and using social media platforms like LinkedIn and YouTube to learn about healthy financial attitudes and personal financial management. It is also important to know that personal financial knowledge is first an attitude that translates to behavior and then becomes a skill with practice (Housel, 2020).

With these steps in mind, people can master the attitude and skills necessary to manage their personal finances. In turn, they can successfully develop into productive members of society; become less burdensome on the government through declining social grants; become better savers and investors; and become more knowledgeable consumers especially concerning debt.

At a macro level, financial literacy balances the consumerist tendencies of modern society through innovation and production – which increases Gross Domestic Product (GDP) – raises national wealth, and contributes to the general well-being of society. Money touches nearly everything and is understandably one of the major causes of stress for adults. Having little or no understanding of personal financial control and management adversely affects the health and well-being of individuals, especially in economically uncertain periods like the COVID-19 outbreak (Mbukanma et al., 2020).

Conclusion

Knowledge of financial skills significantly facilitates building and sustaining wealth. Failure of stakeholders, including the government, financial institutions, and the general public, will impede national and personal growth and contribute to the loss of a generation of financial literate persons, while increasing exposure to avoidable financial risks. Hence, it is relevant for financial literacy to become commonplace as information technology for the purpose of maintaining sustainable financial stability and a better future financial well-being for all. In 2022, the economic prediction for global recovery from the effect of COVID-19 was bleak, due to the effect of the Russian-Ukrainian conflict on energy supply, food availability, climate change, and slowing economic growth. The time for the inclusion of personal financial management in the educational curriculum is now.

Finally, personal financial literacy offers a straightforward mechanism for personal financial success and rational financial decisions. Surviving financial strain during periods of financial stress, like the COVID-19 catastrophe that may follow, is dependent on sound personal financial literacy, which the educational system provides. In the meantime, there is also a place for personal initiative that should include personal finance learning and an unceasing personal commitment to acquiring needed financial knowledge, given the ubiquity of technologically-aided learning platforms available online and from social enterprise.

Reference list

Aliche, T. (2021). *Get good with money: Ten simple steps to becoming financially whole.* Rodale Publishers.

Central Bank of Nigeria. (2015). *Nigeria financial literacy baseline survey report.* Central Bank of Nigeria.

Drentea, P. (2000). Age, debt and anxiety. *Journal of Health and Social Behavior,* 41(4), 437–450.

Herman, M., & Bala, M. (2007). Personal finance education: An early start to a secure future. *CTMS Journal,* 3(1), 39–46.

Housel, M. (2020). *The psychology of money: Timeless lesson on wealth, greed, and happiness.* Harriman House Publishers.

Lusardi, A., & Mitchell, O. S. (2014). The economic importance of financial literacy: Theory and evidence. *Journal of Economic Literature,* 52(1), 5–44.

Mbukanma, I., Rena, R., & Ifeanyichukwu, L. U. (2020). Surviving personal financial strain amid COVID-19 outbreak: A conceptual review of South African context.

AUDOE, *16*(6), 175–190. https://dj.univ-danubius.ro/index.php/AUDOE/article/view/536/1020

Nigerian Educational Research and Development Council (NERDC). (2008). *Frequently asked question (FAQ): The new 9-year basic education curriculum, structure*. Federal Republic of Nigeria.

Turner, T. (2022, April 8). *Personal finance*. www.annuity.org/personal-finance/

25

SELF-COACHING FOR PANDEMIC SURVIVORS

Vikram Kapoor

When the world was turned upside down

Head down, eyes fixed on the floor in the cramped corridor of the plane, I shuddered at the thought that there might be a bomb on board. Maybe I was the bomb, or the passenger to the left of me. We were leaving Istanbul, Turkey, with reports of infection growing in neighboring countries. *Gosh this lovable wife of mine just spent the entire day as a tourist in the old spice market.* No one was wearing masks until that last day, and then suddenly there was a mad dash for them. I was wearing a bug-like 3M mask now, which cost me over $10.00, with a plastic release valve for air – I feel like I'm in a bad '80s movie and it just won't end. *How do you even eat with a mask on?*

Shortly after I joined the United Nations in October 2019, I began planning a multi-country fieldwork assignment ("Mission"), to begin in February 2020. Of course that proved to be a challenging moment in time for the world, and I arrived in Nairobi, Kenya, to learn of the COVID-19 outbreak in Milan the same day.

This was a particularly hairy problem for me because Italy was to be the third stop on this Mission trip. We had people planning to come in from all over the world to the UN's famous training center in Brindisi, Italy. I had preemptively messaged our colleagues from Asia to ask that they reconsider their participation for everyone's health and safety – a move that felt both sadly exclusionary and cautiously optimistic.

We indeed were able to finish our training program in Nairobi and Istanbul (having had to cancel the one in Italy), but it was a nail-biter for many participants who rushed home before lockdowns. I think our Sri Lankan colleagues, who requested and received a special waiver to attend the training, then found

DOI: 10.4324/9781032690278-29
This Chapter has been made available Under a CC-BY-SA license

themselves racing out of Istanbul to barely make it home before the mandatory isolation took effect. I admire their bravery and commitment to learning in person, and I also appreciated the hard work of our colleagues from Sudan, South Africa, Ethiopia, North Macedonia, Jordan, and so many others who interacted in these coach trainings. There was indeed a great deal of sharing and learning as we broke bread together, learned cultural sayings from each other, and dug into key narratives we had about advice giving, problem-solving, and conflict management.

In the weeks that followed, my teammates and I developed online instruction to bring in many more people into the fold through remote training. This last point bears repeating. I personally was involved in training more than 75 coaches all over the world remotely, in an unprecedented move for the United Nations.

In fact, the pandemic must have accelerated the UN's progress in virtual and remote work areas by years, if not decades. Suddenly, colleagues in far-flung places were tasked with figuring out how to have a fully remote workforce, and by and large, people stepped up to the challenge in remarkable ways. I had a good time talking to folks who were clearly on their phones *en route* somewhere, with all kinds of exotic sounds in the background.

I wonder if the UN and international organizations will continue to evolve in the areas of technological connectivity and virtual collaboration. All indications are that they will. I also see this same IT revolution in the facilitation, coaching, and mediation space, where people can suddenly leverage technology to reach people more easily, and in ways that allow for more psychological safety and autonomy than ever before.

We are all survivors

Plainly put, we always have some choice on whether to be a victim or a survivor. Are we letting things happen to us, or are we exerting some influence on how things happen, even if that influence is limited to our own feelings and beliefs? Think of the famous psychoanalyst and Auschwitz survivor, Viktor Frankl (who founded logotherapy, which describes a search for life's meaning as a central motivational force), and any number of other public trauma survivors. Believe it or not, humans have enormous capacity for compassion and self-compassion. Do we have the requisite empathy for ourselves and others?

We are all survivors of one of the worst crises the world has seen in over 100 years. Maybe 500 years, or maybe an all-time worst, if you count actual losses of human potential. Just imagine the unhappiness, the broken relationships, the uncertainty and fear, and the loss of life and livelihood. To me, it does not seem to help if we dwell on those details now. The point from a self-coaching angle is about mindset. Can we flip our mindset if we get stuck in victim thinking? As survivors, we have so much to be grateful for.

Some self-coaching tools to help us

As you and your team, organization, and community continue to grow and adapt, there are some cutting-edge self-coaching tools that can help you, just as they helped me and hundreds of my clients and colleagues on our journeys through the pandemic.

What is self-coaching?

Self-coaching is merely an approach, centered on the idea that you can guide yourself through a series of exercises and activities to unlock new ideas and possibilities. Self-coaching is effective through the use of a journal, a voice recorder, a mirror, and/or a cellular phone, or we may naturally gravitate to a particular instrument.[1] What actually works for you?

How to use self-coaching

I can give you one technique using each tool. Say you want to work on your distractions during the pandemic – you are easily distracted, anxious, and you want to build more focus and be "in distractable." How would you go about doing this? Here is a potential approach:

1. **Positive priming.** You might utilize a little mindfulness in the morning or evening, or if you prefer, you could creatively daydream (creative visualization) about a positive future. Some gentle positive priming to get you in the right space includes slow diaphragmatic breathing and long exhales (which activate the vagus nerve to regulate breathing and heart rate).
2. **Journaling or using a voice recorder** might be a nice follow-on there. You could keep track of your emotions related to distractions and the pandemic. You might speak to what these emotions say about you, and how they fit in the larger context of your behavior and life. Expressive journaling about emotions can be a very powerful technique to explore your thoughts, and using a voice recorder can also expedite things if you are "on the go."
3. **Leverage your strengths and superpowers**, which you can learn through even simple assessments like the VIA – virtues in action, which is free, and the *Gallup Strengthsfinder* (requires a fee), both of which are easily searchable on the internet. Once you have a sense of your strengths and core values, you can utilize those to build a "personal mission statement" that inspires you and reminds you of your purpose, thus helping you rebound from distractions. For example, you might use your top strength(s) to serve a certain population or cause, because of a particular core value. For example, "I use my top strength of Learner to serve young leaders and change makers because I deeply value developing the next generation." What is

your personal mission statement? If you need help with strengths and values, there are a number of free worksheets online you can use. You can also ask your friends and family for their thoughts. They may know something about you that you are missing.
4. **Supplement your own breathing and meditation with your phone, using an app or other guided meditations.** There are some wonderful (free) self-compassion meditations available if you search for "Kristin Neff and self-compassion meditations" online, for example. I recommend these because any behavioral change takes a lot of work, and it can help to have a little compassion for yourself.
5. **Observe positive changes and celebrate those in front of a mirror and reward yourself.** As we do good things, the behavior is cemented when we also trigger the Reward Center of the brain. We can do this through small celebrations, rituals, new experiences, and ever more gratitude.

In so doing, perhaps you would find a range of new ways to address the issue of distractions. This can work for any number of related matters, such as childcare or eldercare during the pandemic, new demands from work, and so forth.

Some may really benefit from using the mirror or a voice recorder. For example, Roman, a founder of a small artificial intelligence startup organization, has been building his brand with resilience, even as others are losing hope and closing down their businesses. Roman meditates and reflects regularly, using the mirror to evaluate his confidence level (i.e., checking to see how confident his words land before a big investor pitch). He finds ways to clear the mental clutter through audio journaling and to reduce self-doubt upon waking. At the time of writing, Roman is enjoying his time in Portugal and is finding opportunities to be of value in the startup space there. He is feeling good about his courage and commitment. All of these practices, when layered to fit one's needs, can be remarkably beneficial with little to no cost.

Another example is Rishi, who was perpetually fearful about the future. During the pandemic, he had become isolated from his friends, stayed home and ordered delivery food, and cheered himself up by working hard and otherwise losing himself in frivolous television shows. How might he turn this around? Rishi began with some basic journaling about the day – from an emotional and energy standpoint. When did he have positive emotions and high energy, and so forth. This helped to create a baseline that Rishi could then decide to shift away from, if he so chose.

He saw that his morning and night routines were zapping him of all energy, so he could make small changes there. In also doing a basic life assessment known as the Wheel of Life, he was able to quickly see that his priorities did not match up with his vision for the life he wanted, mainly in that he was not investing the time into his relationships with family and friends. In this way,

Rishi was able to coach himself to pivot early in the pandemic, stop languishing, and make the most of the time, in his case, by reconnecting with friends all over the world.

Closing thoughts

There is no question that the pandemic will continue to impact our well-being, our personal development, and our quality of life in unequal ways. Income inequality for you will increase if you did not own a home before 2022. The disparity will be stark between those who must report to wage grade jobs each day, as opposed to those in the knowledge economy who can work remotely, thus saving money and time on commute, maintenance of work clothes, and so forth. Those who work remotely may also be subject to distance bias, causing them to miss out on potential income over their lifetimes. We look out at a whole new world, and there is much uncertainty to be sure.

The pandemic caused untold misery for those who acquired COVID-19 and suffered the consequences, or others who were frustrated and fearful of the future. Under conditions of isolation caused by quarantines and lockdowns, emotional stress increased, including higher incidence of anxiety and burnout.

Many tools and techniques are out there to help us bounce back from setbacks and recover from terrible loss and tragedy. There are people and practices to help us, and when we cannot find or afford the help of a guide, therapist, mentor, or coach, we can take solace in finding or regaining the capacity to coach ourselves. If the pandemic awakens in us this capacity, then perhaps there is hope yet for a better world indeed.

Note

1 See Chapter 21, "How a Gen X became a Gen Z at heart" by Bina Patel.

PART V
Technology and culture

26
INTRODUCTION TO THE CULTURAL SECTOR

Suresh Nanwani and William Loxley

Introduction

Technology for cultural development strengthens society and the individual when ICT adapts itself to human needs that encourage personal growth in knowledge, career, and social relations (Batteau, 2009). Dynamic societies recognize that cultural development improves societal efficiency through information and knowledge flow in art and science (Ryan, 2020; Arpaci et al., 2021). Examples of these are what occurred in Paris of the early 1900s, Argentina in the 1920s, and Lebanon in the 1950s. This recognition leads to daily problem solving and greater awareness of bigger issues such as the meaning of life for individuals and cultures.

The pandemic showed society that excessive time spent passively on the internet for entertainment purposes does not improve the human condition. For example, digitalization should help the world self-improve through health, education, work, trust in public sector planning, blended learning, and work from home in the "zoom world." Yet passive participation online and gaming activities could stunt learning and creativity (Pandya & Lodha, 2021). The mantra of "eat, sleep, game" is not a good one for a dynamic society.

Human development improves through creative ingenuity that relies on open-source networks. ICT provides a pathway to additional dimensions of being human (Cecchetto & MacDonald, 2022). Smartphone digital access to social media becomes the norm as newspapers and library cards become extinct. More and more the internet determines what and how we think. It appears social media is the media for social interaction online, which allows quick affordable access to better communication through instant messaging. As such, culture moves into this new realm of possibility where digital designs

DOI: 10.4324/9781032690278-31
This Chapter has been made available Under a CC-BY-SA license

shape reality. The culture is defined by augmented reality as technology races ahead of human comprehension. Will culture lose its creative punch due to passive relations with internet entertainment? If this happens, humans will be on track to lose their humanity through lack of personal authentic relationships (McMahon, 2020).

Culture plays an important role in structuring the identity of individuals (Cover, 2021). It allows knowledge and wisdom to be distributed among communities through art, philosophy, and science, to create humanism out of concerts, cinema, and public debates. Culture is heavy-laden with education and training goals when preparing populations to meet societal goals and find individual awareness of life's meaning. Consequently, governments have an obligation to restructure education systems to ensure that youth would benefit from schooling. This includes promoting learning that stimulates social networking and discussion (Ryan, 2022).

ICT in the post-modern world is an ideal vehicle to transport ideas and creative flow throughout life. These ideas merge in the form of entertainment as in gaming, but also in personal development that ensures lifelong learning, both formally and informally through concert halls and special interest clubs. The pandemic quickened the pace between ICT innovations and the design of the digital world. While this pace was already fast, the pandemic accelerated it further. It is now moving at lightning speed toward the "internet of everything" in the metaverse. ICT is changing the virtual world of the avatar into a three-dimensional world of the hologram that offers individuals many choices or fantasies to pursue personal identity at the expense of reality. For instance, some individuals will construct their "perfect self" with no flaws by interacting with others in the metaverse to bolster self-image.

ICT affects the demographics of society profoundly (Friedman & Parker, 2021; Zhao et al., 2021). Generation Z is fast becoming plugged into the handheld cell phones of the metaverse, and it remains to be seen what effects digital life will have on their human condition. Will individuals be drawn passively into a world of scroll-click-watch or take an active role in constructing virtual personal worlds using machine learning to perfect their ideas? For example, spending more time listening to music during the pandemic should have helped individual wellness because music soothes the mind and the soul. The outcome will determine the well-being of future generations as humans or robots vie to reconstruct culture.

Seven essays

Technology and cultural development were greatly affected by the COVID-19 pandemic. Here, development refers to cultural activities that raise the human dynamism found in both institutions and individuals. The theme covers social awareness that improves psychological disposition contributing to human knowledge. Major issues include values and behaviors that encourage self-improvement, life-long learning in fashion industries and green living, enterprise creativity, and ICT. Topics found in Theme Four include relations between humans and ICT, AI, internet advances, and trends toward the cultural metaverse. Other topics include cross-fertilization of philosophical exchanges between East and West courtesy of the internet during COVID-19; effects on the creativity of musicians; technological prowess "Happy" Bhutan; and personal satisfaction from closeness to nature.

Reference list

Arpaci, I., Al-Emran, M., Al-Sharifi, M., & Marques, G. (Eds.). (2021). *Emerging technology during the era of COVID-19 pandemic.* Springer.

Batteau, A. (2009). *Technology and culture.* Waveland Press Inc.

Cecchetto, D., & MacDonald, C. (2022). Listening through a pandemic: Silence, noisemaking, and music. In I. Gammel & J. Wang (Eds.), *Creative resilience and COVID-19: Figuring the everyday in a pandemic.* Routledge.

Cover, R. (2021). Identity in the disrupted times of COVID-19: Performativity, crisis, mobility and ethics. *Social Sciences and Humanities,* 4(1), 100175. www.sciencedirect.com/science/article/pii/S2590291121000711

Friedman, E., & Parker, A. (2021, April 12). *An early look at the impact of the COVID-19 pandemic on demographic trends.* The Rand Blog. www.rand.org/blog/2021/04/an-early-look-at-the-impact-of-the-covid-19-pandemic.html

McMahon, C. (2020). *Psychological insights for understanding COVID-19 and media and technology.* Routledge.

Pandya, A., & Lodha, P. (2021). Social connectedness, excessive screen time during COVID-19 and mental health: A review of current evidence. *Frontiers Human Dynamics.* www.frontiersin.org/articles/10.3389/fhumd.2021.684137/full

Ryan, J. M. (Ed.). (2020). *COVID-19 volume II: Social consequences and cultural adaptations.* Routledge.

Ryan, J. M. (Ed.). (2022). *COVID-19: Cultural change and institutional adaptations.* Routledge.

Zhao, J., Ahmad, Z., Khosa, S. K., Yusuf, M., Alamri, O. A., & Mohamed, M. S. (2021). The Role of technology in COVID-19 pandemic management and its financial impact. *Complexity, 2021,* 1–12. www.hindawi.com/journals/complexity/2021/4860704/

27
WILL TECHNOLOGY REPLACE OR RECREATE HUMANS?

Arthur Luna

Introduction

Information technology and health organizations have experienced increased demand due to the COVID-19 pandemic. IT support for health workers is needed to help individuals continue with mask wearing and social distancing in many urban areas and public settings. With over 500 million citizens living in the five largest countries in Southeast Asia (Indonesia, Malaysia, Philippines, Thailand, and Vietnam), nearly 200 million contracted COVID-19 and 350,000 died (WHO, n.d.). Since the region comprises many islands in rural and remote settings, this data mostly reflects big and medium urban areas where data reporting is possible.

The government addresses policies and provides information about the pandemic through social media, texts, emails, and television. It also encourages the public to use advanced technology to stay in touch during the quarantine and lockdown periods when people stayed indoors and online. During lockdown, people texted, audio-video chatted, watched movies and videos, played games, shopped online, and studied using distance education programs. In this chapter, I talk about how technology innovations changed the way individuals used the internet and social media. Younger generations became especially connected to cell phones, tablets, and computers, spending much time online during lockdown and coping with virtual face-to-face contact on the internet.

In addition, technological progress affected banks, schools, hospitals, and other public and private businesses during work from home (WFH). The IT platforms developed and marketed by large companies and small-scale entrepreneurs increased during Covid times, as people spent more time working and entertaining themselves through learning and earning a living online. Life online has changed the way people ordered food, shopped, transacted banking matters, sought information and social services, and interacted with others on online platforms. Advanced technology makes this all possible, including IT developers that provide applications and company systems that allow work from home.

Technology platforms

Technology has accelerated digital and contactless interactions among many large corporate players that were expanding their products and services around the world even before COVID-19 struck. Facebook, Twitter, Instagram, and WeChat are some of the technology platforms that grew into hundreds of millions of users. Using these platforms kept people in touch with relatives, friends, coworkers, and also served as social media devices to keep people informed of news and mass culture events.

Other platforms like TikTok included systems that allowed audio and video to entertain and carry on cultural activities in the arts and sciences. In schooling, business, and government, virtual meetings occurred through Zoom, TeamViewer, BlueJeans, and Microsoft Teams. Many companies offered services through online transactions like banking, bills payment, shopping, transport services, and other activities to simplify life under lockdown in almost all developed urban regions of the world.

I work in Banco de Oro Unibank (BDO) in IT support. My company's management provided necessary platforms of technology, laptops, and internet access through VPN (virtual private network) for employees to work from home using voice over internet protocols. Staff access was given to company portals on mobile phones secure information and communication, which could be updated in company contingency plans.

Society is now only getting back to normal after three years of the pandemic, which has severely disrupted life due to the volume of affected individuals and fewer earnings for workers and families. In some cases, establishments did not survive due to lockdowns. Classroom education was shut down, and no universal alternative has yet been found to provide distance learning. Hospitals were overwhelmed from the weight of COVID-19 patients admitted to ICUs and long-term hospital care. This once-in-a-lifetime event shook nations, communities, families, and individuals to the core as everyone wondered what would happen next. As a bank employee and IT expert, I have the opportunity to serve people who need application platforms like BDO apps, websites, policies, and process.

Who provides the technology platforms?

If technology platforms are developed and marketed by outsiders in the areas of online learning, work, social media, and data streaming, will societies lose control to large corporations and their decisions to manage and monitor news and information? It is a big challenge for governments and companies to secure privacy of information coming from working at home while using IT platforms.[1] Many new technologies come from advanced countries with their many advantages while lower- and middle-income countries still lack both the quantity and quality of technology to develop new technologies.

Public monitoring has also changed with the arrival of COVID-19 and the social and economic shutdown of social life. In the banking industry, work from home has benefited most employees especially those living in urban areas. In some companies, the move to virtual learning has not benefited workers, especially those living in remote and poor areas. Educators and students still struggle to find workable learning models to replace or supplement classroom education (Richards, 2010).[2] Public health organizations continuously focus on vaccinations and frontline support for health workers in a health system that is poorly designed and funded.

Newscasters and broadcasters are continuously providing information about the pandemic, economy, and the status of nations using traditional applications and technology platforms that sometimes do not encourage viewing. To counter problems, governments and many companies offered online classes and trainings using advance technology platforms to improve individuals' distance learning. The business community had to accept WFH and that innovation has altered the emotional well-being of many young educated workers who do not want to return to the office. This large issue has affected the way businesses redesign the workplace and provide incentives to keep talent from leaving if the workers are unhappy.[3]

Finally, technology has accelerated the change in social networking on the internet when individuals are still overly reliant on time spent online. All of these major changes require new technologies to address the way societies interact on social media platforms. While businesses may worry that their human resources may not keep up with machine learning, middle-income countries worry about losing control of information and data flow between public- and private-managed organizations. This issue is important for public policy when determining control of information.

Technology requires data

The aforementioned issues require big data algorithms and internet providers to manage information when monitoring outcomes. Most middle-income nations have only cobbled together existing basic data collection sources on health cases vaccinations and public assistance, and the results are not highly reliable.[4] In education, hardware infrastructure is lacking nationwide and limited exposure to "off-the-shelf" software materials is often not useful under local conditions (Mir, 2014). Monitoring of public health outcomes still lacks reliability in finding trends. In the business world, companies redesigned data systems to connect WFH, internal and external communication, networking, and AI interventions where feasible.

In media, because everyone living indoors spent time on the internet, much effort has been spent improving the quality of entertainment, like video-watching and following the social media lives of influencers, shoppers, and mass culture followers. As a result of this increase in time spent online, technology now holds its grip on digital media with its potential for positive and negative influence. Potential for good includes intellectual and emotional growth, while negative influences include psychological bullying, hacking, and spreading fake news.

When COVID-19 finally fades away, it will be easier to see exactly how powerfully technology changed the world. Will humankind overrely on IT-based applications and algorithms developed especially by machines? Will IT become a product created and controlled by humans or machines? These questions will require careful thought and examination because it is important for future generations to retain control over technology. We know technology will impact the environment and automation greatly, but it will also influence many human processes like education and health.

People take for granted that changes in schooling, public service delivery, the workforce, and the spread of cultural ideas will naturally occur, but they do not know by how much and in what way (Goldman & Katz, 2010). The 21st century is in danger of creating two worlds side by side but far apart in technological capability. Advanced nations with their wealth and talent continue to innovate rapidly. On the other hand, developing societies with limited resources and shortages of scientific talent are forced to adapt "off-the-shelf" technology from others that limit local culture and ideas from flourishing.

Southeast Asia prospects

While Southeast Asia is progressing along economic, social, and technological lines, COVID-19 accelerated the pace of technical innovation that has prevented the region from keeping momentum. Likewise, COVID-19 has retarded educational learning among the majority, especially the poor. Public health also remains woefully inadequate to meet national needs, especially among rural populations. Both governments and businesses are benefiting greatly from recruiting talent from local higher education while borrowing from best practices worldwide. This is especially true in the business world but also among the young technocrats in government. These technology skills and good business practices learned in universities and put into practice allow government and business to adapt to change caused by the pandemic.

In technical enterprise, the social media world of the internet, web, and augmented reality seems to be moving forward while some bad actors continue to hack and spread fake information and news for personal advantage. Government regulations should expand to meet this threat, while technologists stay ahead of and work closely with artificial intelligence to remain in charge of directing cultural change.

I see the need for innovative technology to make individuals and companies more productive. In banking practices, employees and clients need to understand application platforms, like access to company portals while working from home. Now, employees can access Microsoft Outlook emails and have virtual meetings via Microsoft Teams. Companies are also using remote assistance and remote desktop connections to provide technical support when there is a problem with device access.

My experience in the pandemic also covers working in the provinces, when BDO set up new branches in Aroroy municipality in Masbate and Roxas municipality in Oriental Mindoro. The latest technical advances and data compatibility helped me to set up the company's IT system. I helped set up equipment and software work smoothly thanks to technical advances and data compatibility. Soon I will work on establishing branches in Palawan, an archipelagic province of the Philippines. Prior to the pandemic, when I worked overseas like in Hong Kong, I found my experience rewarding in terms of installing advanced technology and reliable internet connection when upgrading applications and IT equipment.

Elsewhere in the United States, Canada, and the Middle East, I found remote support can easily isolate and troubleshoot technical problems. In these areas, IT is very advanced (Soper, 2014). In due course, I am hopeful that IT in the Philippines and neighboring countries in Southeast Asia will considerably improve.

In terms of cybersecurity to protect internet-connected systems such as hardware, software, and data, companies are now using Symantec Endpoint Protection developed by Broadcom Inc. It is a security software that consists of anti-malware, intrusion, prevention, and firewall features for servers, desktops, and laptop computers. I find this software security useful in the pandemic times to prevent sensitive information from falling into the wrong hands.

Conclusion

Future society will depend on technology and internet data to make everyone more authentic in the real and online world. Authenticity allows individuals to become the best they can be, and technology is key to helping in the process. Individuals are obliged to be more knowledgeable in technology for society to advance. This is especially true for Generation Z not to be left behind in the digital world (Bellanca & Brandt, 2010).

To accomplish this, education systems are required to make the learning student-centered for everyone. It requires using big data policy to serve the public good and improve business practices, and social networks that produce art and human well-being through creative experiences online and offline. If technology can make the networking world of human interaction easy to access and learn from apps, then it is possible that future generations will not be replaced by machines and technology (Russell, 2019). Rather, humans will be reinvigorated and thrive in the world of technology through creating a meaningful culture.

Notes

1 See Chapter 28, "Privacy issues in online education technologies in China" by Li Mengxuan.
2 See Chapter 9, "COVID-19 and tertiary education: Experiences in Lesotho institutions" by Tsotang Tsietsi.
3 See Chapter 21, "How a Gen X became a Gen Z at heart" by Bina Patel.
4 See Chapter 4, "Digital technology: a best friend for implementing COVID-19 policy in China" by Li Xudong.

Reference list

Bellanca, J., & Brandt, R. (2010). *21st century skills: Rethinking how students learn.* Solution Tree Press.
Goldman, C., & Katz, L. (2010). *The race between education and technology.* Belknap Press of Harvard University Press.
Mir, N. F. (2014). *Computers and communication networks.* Pearson College Div.
Richards, W. (2010). *Blogs, wikis, podcasts, and other powerful web tools for classrooms.* Corwin Books, Sage Publishing.
Russell, S. (2019). *Human compatible: Artificial intelligence and the problem of control.* Penguin Books.
Soper, M. E. (2014). *The PC and gadget help desk: A do-it-yourself guide to troubleshooting and repairing.* Que Pub.
World Health Organization. (n.d.). *WHO Coronavirus (COVID-19) dashboard.* World Health Organization. Retrieved February 4, 2023, from https://covid19.who.int/table

28
PRIVACY ISSUES IN ONLINE EDUCATION TECHNOLOGIES IN CHINA

Li Mengxuan

Introduction

No further explanation is needed as to how school work has been upended by COVID-19. I was in my second year of undergraduate and happily planning for my graduate school exchange to Germany. But suddenly schools were all shut down around the world. Luckily, or maybe unluckily, all parties have reacted rapidly to put students back in "school" with Education Technologies (EdTech) websites and apps. But online education is drastically different from traditional education. In the rush to connect students to virtual classrooms with little adaptation time, problems have risen. One of these is the failure to protect the privacy of students and teachers because the internet environment is too easy for surveillance.

This topic is especially important as we enter the digital era. Tons of new technologies are coming out – fintech, health tech, insurance tech, etc. Personal information in the digital era has become a precious storehouse of value before we realized it. Now the question is more urgent than ever – technology companies, and even governments, try to use personal data and are often blind to privacy collection issues in these technologies. In discussing privacy issues in online education, we can also cast insight into privacy issues in other areas.

In this chapter, while I will touch on both higher education and compulsory education, I will focus heavily on higher education. It is important to notice these two are very different in many aspects. To complete this essay, I have talked with my cohort and my father, who has served as a middle school math teacher in Guangdong province for over 30 years. From my conversations with them, I have built a more comprehensive perspective on this issue. In this chapter, I will explore how EdTech is infringing on our privacy. Based

DOI: 10.4324/9781032690278-33
This Chapter has been made available Under a CC-BY-NC-ND license.

on that, I will introduce some of the current and potential solutions to these problems and analyze their strengths and weakness to see how we can learn from them in the future. In conclusion, I will discuss some feasible activities to ensure online education in China will grow positively.

Emerging education technologies and related privacy issues

Many EdTech products endorsed by governments and used by students during COVID-19 school closures were found to harvest data unnecessarily and disproportionately for purposes unrelated to their education. Most EdTech companies did not permit students to decline being tracked; most of this monitoring happened secretly, without the students' knowledge or consent. In most instances, users couldn't opt out of such surveillance and data collection without opting out of compulsory education and giving up on formal learning altogether during the pandemic.

Types of online education and emerging EdTech involved

To get into specific privacy issues arising from online education, I will introduce typical EdTech activities in different types of online education, with the help of its real operation pictures on the website pages. Based on my own experience in both Chinese and U.S. law schools, I will outline an EdTech program I used in China in comparison to its international counterparts.

Synchronous Learning: Tencent Meeting versus Zoom

Tencent Meeting, also known internationally as VooV Meeting, is the Chinese counterpart of Zoom. Tencent Meeting and Zoom are both online meeting platforms, mostly used in live lectures in online education. They apply a similar business mode, where individual users have limited free access to the meeting (e.g., limited capacity or time) (Amo et al., 2019).

Asynchronous Learning: Chinese University MOOC versus Coursera

Massive online open course (MOOC) is a popular form of asynchronous learning. The Chinese University MOOC is a government-based asynchronous online education platform. It provides the public with over 1,000 MOOC courses from over 900 well-known Chinese universities. When enrolled in a course, participants can watch videos, participate in discussions, and submit assignments, interspersed questions, and the ultimate exam. Students are awarded an ID-verified certificate on successful completion of a course, which includes a certificate number and a QR code for verification. Courses offered by universities are mostly free, but those offered by private companies might require payment.

Coursera is a global online learning platform that offers online courses and degrees from universities and companies around the world. As a private company, Coursera is more expensive in contrast to Chinese University MOOC. For example, there are two types of subscriptions to Coursera Plus: monthly (USD 59 a month) and annual (USD 399 a year).

Collaborative Learning: Padlet and Canvas

Padlet is not unique to China. It is a U.S.-based EdTech start-up. It is a real-time collaborative web platform where users can upload, organize, and share content to virtual bulletin boards.

Canvas is a learning management system largely used in the U.S. Many states across the U.S. have confirmed a partnership with Canvas's developing company Instructure Inc. to adopt its platform for their educational institutions. The platform functions using engaging course content, quizzes, grades, data, insights, students' interaction with educators and peers, etc.

Privacy issues in China online education

Why does privacy intrusion in online education need to be a specific focus? The answer lies in the privacy characteristics of EdTech users. Firstly, users' online privacy on EdTech varies, including not only information such as a home address, parents' occupation, income, education, etc., but also students' daily subject exercises, tests, test scores, attendance, discipline, and on and on. Secondly, in addition to the aforementioned information, explicit private information such as families and individuals, students' learning habits, learning ability, personal characteristics such as hobbies, learning paths, and personality traits can also be inferred. Once this invisible private information is leaked, it will have a greater impact and potentially harm students' freedom of thought.

Last but not the least, students cannot decide their right to privacy. Young students do not have legal independence, and their privacy protection is usually done by parents or teachers. At the same time, few students and parents are asked for consent to use certain EdTech. Parents and teachers operated on blind faith that their governments would protect their rights when providing education online during COVID-19 school closures.

There are countless types of privacy involved in EdTech, and there are still new types arising all the time. Here I will only analyze the privacy violations that have occurred in the EdTech I introduced earlier (Managed Methods, 2022).

i. **Synchronous teaching.** For webcasting technologies used in synchronous teaching, its usage normally assumes personal data of schools, universities, and training institutions, such as registration IDs including personal

accounts and PINs, personal information, lists of participants, video scripts, and other relevant information upon requirement.
ii. **Asynchronous learning.** When a MOOC course has been selected for online learning, one needs to register on a MOOC platform. In this context, logged personal data typically contains basic personal information and track records.
iii. **Collaborative learning.** Online Learning Management Systems presents a data-rich environment where there is extensive learning and teaching data produced by students and teachers with each online learning session. It is important to secure personal data collected or produced from students and teachers, including personal information, learning content, discussion scripts, learning records, performance data, etc.

Analyzing online privacy issues under social frameworks

Solutions in China under a social framework

Under the social framework, online education marries two of the hottest sectors in China – education and the internet. Although one may not immediately link the two sectors, they do share a lot in common. Both are strongly supported by the PRC authorities but are also subject to increased scrutiny. In addition, education and technology closely touch the lives of most Chinese consumers.

Many will say that privacy only came into the public eye in recent years in China. There is a lack of social organization or industry associations to regulate at the social level. But parents' associations at schools are beginning to take an active role in protecting students' rights. One good example at the governmental level happened in my hometown. Guangdong Province Department of Education adopted a positive list policy regarding online education. As early as July 2019, Guangdong introduced the policy that EdTech for primary and secondary school students must be included in the whitelist of products inside the library, selected after the authority's review and registration, before recommending them for students' use.

Potential approach to improve information flow

The value of online education should receive further attention. "Ensure inclusive and equitable quality education and promoting lifelong learning opportunities for all" is the goal of the Education Agenda for 2030 Sustainable Development (United Nations, n.d.). Online learning is the foundation for achieving this educational goal, and it applies not only to education in times of emergency but also to education in the future. Meanwhile, data security and personal privacy protection are urgent. The basic knowledge of personal data protection in the process of online learning, such as setting up devices, registering online learning platforms, and learning through them, is of great importance for personal data security.

To promote the protection of personal privacy in online learning, the government's policy standards, the industry's technical guarantee system, and the behavior of other stakeholders should all work together to create a secure environment for learning.

i. **Students, parents, and schools.** EdTech users should improve self-practices on data protection. In the meantime, society should raise awareness of privacy protection. Even when supporting education, it should prioritize learning over anything else.
ii. **Administrators.** Remedies are urgently needed for users whose data were collected during the pandemic and remains at risk of misuse and exploitation. Governments should conduct data privacy audits of the EdTech endorsed for learning during the pandemic, remove those that fail these audits, and immediately notify and guide affected schools, teachers, parents, and students to prevent further collection and misuse of data.
iii. **Relevant companies.** In line with child data protection principles and corporations' human rights responsibilities as outlined in the *United Nations Guiding Principles on Business and Human Rights*, EdTech and advertising technology companies should not collect and process users' data for advertising purposes. Companies should inventory and identify all data gathered during the pandemic and ensure that they do not process, share, or use data for purposes unrelated to the provision of education. Advertising technology companies should immediately delete any data they received. EdTech companies should work with governments to define clear retention and deletion rules for data collected during the pandemic.
iv. **Engagement.** Education administrations should provide training programs for the ministry staff, teachers, and other school staff in digital literacy skills and protection of student data privacy needed to support teachers who conduct online learning for children's safety. Last but definitely not the least, administrators and others should seek out students' views when developing policies that protect the best interests of all in online educational settings. They should also meaningfully engage children in enhancing the positive benefits that access to the internet and educational technologies can provide for their education, skills, and opportunities.

Conclusion and prediction

Schools are getting used to EdTech and liking it so far. Even when we meet in person now in U.S. law school, schools still use EdTech to facilitate accessibility to course materials, office hours, and so on. In the post-pandemic era, it seems our reliance on digital services that enable online education will only increase.

In the meantime, EdTech is developing with more and more cutting-edge technologies. With this, technology might touch on more users' privacy, some of whom might not even be aware of it. For example, virtual reality and artificial reality technologies are becoming popular in education. Their potential gathering of highly personal biometric data, such as iris or retina scans, fingerprints and handprints, face geometry, and voiceprints, will continue to be a concern.

The good news is that public awareness of privacy rights is growing, even in China. Big EdTech companies realize they cannot free themselves from privacy issues while making money from EdTech, because privacy issues affect users' trust. Privacy protection is becoming a decisive element in the future development of EdTech and all tech industries. Online shopping, with a relatively long history in China, is also facing the same set of issues.

It all seems sensible on the whole, but I will also mention my doubts about monitoring in Beijing. EdTech is still facing uncertainty in China due to the unpredictable policy of control applied to new technologies. There are two recent landmark actions of the government that are casting doubt on the development of EdTech in China. One is in July 2022, when the Cyberspace Administration of China fined Didi Global Co., the Chinese ride-hailing giant, accusing it of infringing national security by misusing users' data.

The other was in July 2021, when the Department of Education implemented the policy to suspend extracurricular tutoring courses in compulsory education, claiming to level the playfield and promote student wellness. As far as I can see, the government is imposing stricter restrictions on private companies while at the same time may not be willing to prioritize regulating itself on privacy issues.

A bright future in EdTech

In the long run, EdTech is a future we all have to face, whether we like it or not. In the broader picture of China and the rest of the world, I can see stakeholders working toward a bright future in EdTech. Yet there is still a long way to go for improvements in online education for sustainable development.

Reference list

Amo, D., Fonseca, D., Alier, M., García-Peñalvo, F. J., Casañ, M. J., & Alsina, M. (2019). Personal data broker: A solution to assure data privacy in EdTech. In P. Zaphiris & A. Ioannou (Eds.), *Learning and collaboration technologies: Designing learning experiences: 6th international conference, LCT 2019* (pp. 3–14). Springer. https://doi.org/10.1007/978-3-030-21814-0_1

Managed Methods. (2022, February 3). *5 EdTech data privacy risks and how they impact student data privacy and security*. https://managedmethods.com/blog/5-edtech-data-privacy-risks-and-how-they-impact-student-data-privacy-security/

United Nations. (n.d.). *Sustainable development goals: Education*. Retrieved February 4, 2022, from https://sdgs.un.org/goals/goal4

29
DIGITAL TECHNOLOGY DURING COVID-19 IN GLOBAL LIVING EDUCATIONAL THEORY RESEARCH

Jack Whitehead

Introduction

We are all influenced by the sociohistorical and sociocultural contexts of where we live and work. Between January 1, 2020, and August 31, 2022, we have experienced the influences of COVID-19 in our different contexts. My experiences in education over the past two years have been influenced by my economic security. While millions of people have been adversely affected by COVID-19, I have benefited from the many invitations to contribute keynotes virtually to webinars and to virtually present accepted paper proposals at national and international conferences.

I cannot overemphasize the economic influence of being able to present virtually during COVID-19. For example, to present face to face at both the 2021 and 2022 conferences of the American Educational Research Association with flights, fees, accommodation, and subsistence would have cost me personally some GBP 6,000. The registration for virtual presentation was no less than GBP 200. Both symposia involved participants from the United States, Canada, India, Nepal, and the United Kingdom, who, like myself, were able to present virtually during COVID-19 while face-to-face presentations would have been financially impossible.

Virtual platforms and digital technology

COVID-19 has moved much higher education and many national and international conference presentations onto virtual platforms. This has enabled the gathering of digital visual data from a wide range of cultural and international contexts, to be used in evidence and values-based explanations of educational

DOI: 10.4324/9781032690278-34
This Chapter has been made available Under a CC-BY-SA license

influences on learning – that is, to create living-educational-theories. I shall focus on the significance of the technology in gathering and sharing these data, including visual data, for enhancing the flow of global values of humanity such as educational responsibility, equity, and critical reflection. The methods for clarifying and communicating the meanings of these embodied values that we express in our practice include empathetic resonance with digital visual data.

The embodied values are the actual values we express in what we are doing, rather than the value-words we use to describe our values in what are called lexical definitions of meaning. I want to be clear about this difference. Ostensive or demonstrative expressions of meaning include pointing to what we are doing with the help of digital visual data and clarifying the meanings of our values as these emerge in what we are doing. Lexical definitions of meaning are when we define the meanings of the value-words we use, such as freedom, justice, care, and love.

The following data are from symposia at the 2021 and 2022 American Educational Research Association with the themes of educational responsibility and equity; presentations at the 2021 and 2022 meetings of the Network Educational Action Research Ireland on "Critical Reflection in Educational Practice"; the 2022 Educational Studies Association of Ireland on "Why a focus on 'What is educational' matters so much in reconstructing education?"; and the 2022 Higher Education Learning and Teaching Association of South Africa on "Transforming practices in Higher Education through critical reflection."

The analysis of this data, using a method of empathetic resonance, clarifies and communicates the embodied expression of values of humanity. An Eastern Wisdom Tradition together with a Western Critical Theory Knowledge Tradition are included, within the generation of living-educational-theories, in improving educational practices within different international contexts and in making original contributions to educational knowledge with values of human flourishing.

Eastern wisdom and Western critical theory traditions

At the heart of my use of digital technology during COVID-19 was an engagement with individuals and communities that draw on both Eastern Wisdom traditions and Western Critical Theory traditions in enhancing educational influences to learn values of human flourishing. For example, Suresh Nanwani uses technology to make his book on human connections freely available. It contains many images from his teaching experience in Chongqing, China. Here is my abbreviated foreword to his book *Human connections: Teaching experiences in Chongqing, China and beyond*, which embraces a holistic fusion of both Eastern and Western traditions:

> Eastern values and thoughts are embraced in a holistic fusion of both Western and Eastern angles. Eastern values and thoughts include *ikigai* in relation to the meaning and pleasures of life. They include yoga as the unity of

the individual as a system of physical, mental, social, and spiritual development within universal consciousness. The writer uses *ikigai*, yoga, meditation, and *tai chi* as a practice, with techniques such as mindfulness, to build awareness and achieve calmness and stability by weaving positive experiences into the brain and self. He practices *vipassana* as insightful meditation to see things as they really are. *Tai chi* is seen as an art form, embracing the mind, body, and spirit by giving a fresh perspective on life. His perspective also includes Organization Development concepts viewed from a typically Western outlook.

Suresh's originality lies in the form and content of his account that twine together the simplicity, intensity, and probity of his passion for education and teaching with the complexity of asking, researching, and answering his question: "What matters most to me?" and "How do I improve what I am doing?" Suresh's creative story captivates an individual's imagination as to the opportunities that life permits for each person to generate their own story, from a passion to improve their own practice with values of human flourishing.

(Whitehead, 2022, pp. xix–xx)

Before I focus on accessing data from my use of technology from presentations during COVID-19, I want to be clear about my meaning of educational.

Meaning of "educational"

I have been professionally engaged in education since my initial teacher education course (1966–1967) at the University of Newcastle in the United Kingdom. What I mean by "educational" is that it involves learning with values of human flourishing. My inclusion of these values is because not all learning is educational. History is full of examples where human beings have learned to violate these values. Being born in August 1944 might have made me more aware of the importance of enhancing the flow of values of human flourishing. The world was at war, and the gas chambers in Europe were, as a matter of state policy, murdering millions of people who were seen as less than human in the Nazi ideology. The recognition of these crimes against humanity has continuously served to strengthen my own determination to contribute to educational influences in learning, in local, regional, national, and international contexts, with values of human flourishing. Here are the values, with access to the data used as evidence in generating valid, evidence-based explanations of educational influences in learning.

Educational responsibility and equity

Responding to COVID-19 meant that I could, along with others present in symposia at the 2021 and 2022 American Educational Research Association

FIGURE 29.1 American Education Research Association (AERA) 2022 Symposium with participants Jackie Delong, Jack Whitehead, Swaroop Rawal, Michelle Vaughan, and Parbati Dhungana (clockwise from top) on April 22, 2022.

Source: © Jack Whitehead

(AERA), discuss the themes of educational responsibility and equity. In the 2021 symposium, I presented with Jacqueline Delong, Jack Whitehead, Swaroop Rawal, Michelle Vaughan, and Parbati Dhungana (see Figure 29.1). As with Suresh's book on *Human connections: Teaching experiences in Chongqing, China and beyond*, I think the visual image in Figure 29.1 (together with the digital video) is important in communicating the energy-flowing and life-affirming values we express from within our sense of community.[1]

For example, Parbati Dunghana explains in her 2022 presentation the importance of an Eastern Wisdom Tradition in creating her own living-educational-theory. She elaborates that *Ardhanarishvara* signifies "totality that lies beyond duality."

The digital technology I used with the still and video images during COVID-19 enabled the following group to gather data on "Cultivating Equitable Education Systems for the 21st Century in Global Contexts through Living Educational Theory Cultures of Educational Inquiry," as they presented at the 2021 Conference of the American Educational Research Association. The participants from top left moving clockwise are Jacqueline Delong (Canada), Jack Whitehead (the United Kingdom), Shivani Mishra (India), Michelle Vaughan (the United States), and Parbati Dunghana (Nepal) (see Figure 29.2).

We all agreed that this image communicates to us all the embodied life-affirming values of human flourishing we experienced while working and presenting together in expressing a totality that exists beyond dualism.

Critical reflection in educational practice

During the COVID-19 pandemic, I was able to present virtually on "Critical Reflection in Educational Practice" to the Network Educational Action Research Ireland (NEARI) Meeting of April 2, 2022.[2]

Digital technology during COVID-19 in global living educational theory research **247**

FIGURE 29.2   Participants at the 2021 conference of the American Educational Research Association with participants Jackie Delong, Jack Whitehead, Shivani Mishra, Michelle Vaughan, and Parbati Dhungana (clockwise from top).

Source: © Jack Whitehead

Here is the summary of notes I prepared for my virtual presentation:

This presentation builds on the NEARI Meet of the 29th January 2022[3] with its theme of Transforming Practices. Stephen Kemmis provided the keynote with ideas from his book on *Transforming Practices* (Kemmis, 2022). I shall explore the implications of including "educational" in Critical and Creative Reflection in Educational Practice for members of NEARI. While working from a different educational perspective to that offered by Kemmis', I do agree with Kemmis' idea that:

Once education systems and the work of schools are conceptualized principally in systems terms, their essential lifeworld character, their grounding in the everyday life of people's lifeworld is obscured and then ignored. . . . Obscuring and ignoring these lifeworld processes, many PEP (Pedagogy, Action and Praxis) researchers argue, is to obscure and ignore the very substance of the process of education. Neoliberal approaches to educational systems management throw the baby of education out with the bathwater. For me, these neoliberal approaches to educational systems management are policies that promote free-market capitalism, deregulation, and reduction in government spending.

These policies include: choice for parents; per capita funding meaning schools driven by recruitment; competition; League tables; management modeled on business – focusing on "efficient" use of resources and budget maximization; a complex infrastructure of testing; undermining professional autonomy in education; making colleges and universities more vocationally oriented to be responsive to market requirements rather than educational values.

My different "educational" perspective is focused on the generation and sharing of living-educational-theories with values of human flourishing as explanatory principles in explanations of educational influences in learning and as embodied by evaluative standards of judgment. A living-educational-theory is an individual's explanation of their educational influences in their own learning, in the learning of others and in the learning of the social formations within which the practice is located. Such explanations help individuals to answer questions of the kind, "How do you know that your practice has improved?" and "What standards of judgement do you use to justify a claim that your practice has improved?"

My emphasis on critical reflection owes much to the Western Critical Theory tradition that stresses the importance of understanding the influence of the sociohistorical and sociocultural contexts within which the educational practice can be located.[4]

The webinar at the 2022 Higher Education Learning and Teaching Association of South Africa (HELTASA) on "Transforming practices in Higher Education through critical reflection" enabled me to continue to stress the importance of critical reflection:[5]

> With the post-COVID-19 turn in higher education it is imperative that we pause and critically reflect on our educational practices in the last 27 months. . . .. The approach rests on each individual's acceptance of their professional, educational responsibility, to ask, research and answer their question, "How do I improve my professional educational practice in Higher Education with values of human flourishing?" It is shown how transforming educational practices in Higher Education rests on each practitioner accepting their professional educational responsibility to subject their own practice to critical reflection. The critical reflection is focused on being accountable to improving practice and generating educational knowledge with using values of human flourishing. It involves generating and sharing one's own living-poster, creating and sharing one's own living-educational-theory and engaging critically and creatively with the living-educational-theories of others and their contributions to the Educational Journal of Living Theories.

Because of my interest in expressing a totality through an Eastern Wisdom Tradition that goes beyond the dualism in a Western Critical Tradition, I identify with Ubuntu ways of living from Africa, which stress the importance of "I am because we are." (Ubuntu is humanity, a quality that includes essential human virtues.) I share my insights from my pre-COVID-19 keynote to the Higher Education Learning and Teaching Association of South Africa (Whitehead, 2009) and my Inaugural Nelson Mandela lecture at Durban University of Technology (Whitehead, 2011) with the value of Ubuntu in living-educational-theories with values of human flourishing.

Summary

Responses to COVID-19 isolation meant that ideas have been flowing more through digital communications in webinars and virtual conference presentations. These have enabled individuals and communities who engage with both Eastern Wisdom Traditions and Western Critical Traditions to link up their living-educational-theories in generating educational knowledges. They do this by exercising their educational responsibility to continuously improve their educational practices and to generate and share their living-educational-theories as values and evidence-based explanations of their educational influences in learning, contributing to the global educational knowledge base.

While I have personally benefited from the greater virtual opportunities during COVID-19 for communications across global communities with their differing cultural and historical contexts, I do not want to underestimate the damage and suffering around the world caused by COVID-19.

Values of human flourishing

The values of human flourishing of educational responsibility, equity, and critical reflections are closely related to community engagements grounded in societies that are suffering from COVID-19. For example, South African universities enable their academics to engage in community engagements and community-based research that are focused on responding to the effects of COVID-19 with these values of human flourishing.[6] Using digital technology during COVID-19, Global Living Educational Theory Research is enabling the growth of empathetic understanding of how individuals and communities across the world are affected by COVID-19. These communications go beyond this understanding as practical campaigns are developed, grounded in social justice, to ensure the equitable spread of vaccines.

Notes

1 The full video of the 90-minute symposium plus a 15-minute pre-session conversation. https://youtu.be/4h_rRDqIJJ8
2 See Part 1 of "Critical Reflection in Educational Practice" presentation. www.eari.ie/2022/04/22/notes-from-nearimeet-2-april-2022/. The digital technology in Part 2 showed participants how they could access over 50 Living Educational Theory Research doctorates freely available from www.actionresearch.net/living/living.shtml
3 See notes at www.eari.ie/2022/02/08/notes-from-nearimeet-29-january-2022/
4 The face-to-face presentation at the 2022 Educational Studies Association of Ireland on "Why a focus on 'What is educational' matters so much in reconstructing education?" (Whitehead & Huxtable, 2022). www.actionresearch.net/writings/jack/jwmh2022ESAIFINAL.pdf
5 The abstract for my webinar facilitation. www.actionresearch.net/writings/jack/jwheltasa140722.pdf

6 You can access the living-posters of colleagues at North-West University and Nelson Mandela University. www.actionresearch.net/writings/posters/homepage2021.pdf to access the evidence on how colleagues are responding in their communities of practice with their values of human flourishing, including Ubuntu, as they respond to the negative effects of COVID-19.

Reference list

Dunghana, P. (2022). Living educational values for enhancing self-educating strategies for equitable education. In J. Delong, J. Whitehead, P. Dhungana, M. Vaughan, & S. Rawal (Eds.), *Cultivating equitable education systems for the 21st century in global contexts through living educational theory cultures of educational inquiry.* www.actionresearch.net/writings/jack/AERA2022sessionprop.pdf

Kemmis, S. (2022). *Transforming practices: Changing the world with the theory of practice architectures.* Springer.

Nanwani, S. (2022). *Human connections: Teaching experiences in Chongqing, China and beyond.* Suresh Nanwani. www.actionresearch.net/writings/nanwani/nanwaniconnections.pdf

Whitehead, J. (2009, November 27). *A keynote presentation at the higher education learning and teaching association of Southern Africa 2009 on risk and resilience in higher education in improving practice and generating knowledge.* www.actionresearch.net/writings/jack/jwheltasakey09.pdf

Whitehead, J. (2011). *Notes for the inaugural Mandela day lecture on the July 18, 2011 in Durban, South Africa, with a 63-minute video of the presentation.* www.actionresearch.net/writings/jack/jwmandeladay2011.pdf

Whitehead, J. (2022). Foreword. In S. Nanwani (Ed.), *Human connections: Teaching experiences in Chongqing, China and beyond* (pp. xix–xx). Suresh Nanwani. www.actionresearch.net/writings/nanwani/nanwaniconnections.pdf

30
I'M GONNA LET IT SHINE

Local musicians in the Virginia countryside during the pandemic

Lori Udall

Country living and music jams

On any given evening, you can almost hear the sweet bluegrass music wafting through the mountain air of "I'll Fly Away" or "This Little Light of Mine" with rifts on the mandolin or banjo. Music jams are the lifeblood and essence of country living. This is the perfect opportunity to practice playing an instrument with others, learn new songs, meet people, and enjoy camaraderie. My guitar teacher Mark Maggiolo always said, "One jam is worth ten lessons."

A self-taught musician speaks

Mark is a local professional musician and teacher (see Figure 30.1). He is a self-taught musician who plays the guitar, bass, mandolin, banjo, fiddle, and ukulele. Before the pandemic, he had two studios in nearby towns and taught out of his home in Hume, Virginia. He also gigged locally at restaurants, bars, and private venues, and participated in local bluegrass jams including Ronnie Poe's garage in Amissville, the Jam in Unison, and the "last Friday Jam" at Orlean Market in Orlean, Virginia. When COVID-19 hit the U.S. in March 2020, Mark's teaching business was devastated.

"I lost all of my students right away," he mused. "It was a two-tiered issue. Some parents were afraid of having their kids exposed to Covid, and others were worried about economics. Sometimes one or both parents lost or quit jobs, and lessons were a luxury they couldn't afford anymore." Mark tried to start Zoom lessons with mixed success. For the most part, he did not have the internet bandwidth he needed in his country home for sharing and teaching live music. The internet reception was unstable but this got sorted out.

FIGURE 30.1 Mark Maggiolo at home.
Source: © Lori Udall

Mark also lost gigs because many music venues closed over the last two years. If they didn't close, they moved outdoors or reduced the number of patrons. "When venues started opening up again, it seemed like everyone was lowballing musicians. We just couldn't get a good price for a gig. I guess they thought we were desperate," Mark said.

Despite the loss of lessons and health risks, small jams started up again among people who knew each well a few months after the pandemic. Some people wore masks and social distanced. Mark has been lucky because he is reaching retirement age and had saved money for retirement. "Music has been good to me," he said. "I paid off my mortgage and my car from my earnings." Well known for his musical knowledge and teaching talent, Mark has been growing his lessons again. Many of his old students are coming back.

Captain Rich and Michelle

A local guitar-playing duo that also frequented the Orlean Market jams since the early years is The Hobo Mariners, consisting of Captain Rich Coon and Michelle Beall, who have been playing together for ten years (see Figure 30.2). Michelle sings vocals and plays rhythm guitar and Rich plays lead guitar. They divide their time between Hampton Roads Virginia, where they keep their sailboat, and northern Virginia. Years ago, they bought their boat in Long Island and traveled along the east coast to Florida, stopping at marinas and port towns along the way.

"Marinas are a lively place for music," said Rich, who resembles Walt Whitman. "All you have to do is dock and pull out your guitar and soon others will be joining with instruments and fun songs."

Michelle, a fiery redhead whose voice takes over a room, is a singer-songwriter who classifies her songs as "American Roots Country." She has written over 50 songs and says her key influences are Merle Haggard and George Jones. They launched their website (www.thehobomariners.com/) to talk about their music.

Before COVID-19 brought everything to a standstill, they ventured into coastal towns and were invited to play at bars and restaurants. They even ran a marina bar for a time. "We had a blast when we docked at Camp Lejeune, a marine corps base in North Carolina," said Rich. "Before the night was over, there were people in 18 other boats playing, singing, listening, and enjoying cocktails."

The Hobo Mariners began living on their boat in Hampton Roads in 2016 and playing in the region. In 2018, mandolin player Jim Roberts became part of the band. "Before Covid we had gigs three out of every four weeks," said Michelle. "We were rockin'! We played Roanoke, Smith Mountain Lake, Floyd, Killarney. In March of 2020, we were driving back from Killarney Virginia when we started hearing the news about Covid. In no time, all music venues were completely shut down."

FIGURE 30.2 Michelle and Rich Coon performing.
Source: © Michelle Beall

"It was hard-going financially but we used the Covid years to regenerate," said Michelle. "I wrote new songs, learned yoga, lost weight, and tried to improve our food diets. I listened to a lot of other musicians' recordings and worked on vocal control and stamina. I feel that has made me a better musician now."

"We stayed in northern Virginia most of Covid, taking care of a family member. It's still difficult to get gigs," said Rich, "but we are slowly growing and getting back to where we were. Michelle has some great new songs. 'Virginia Lament' is a very special song."

> *And I'm dreaming of Appalachian moonlight,*
> *Longing for sunset on the bay*
> *Looking for a change to come, gotta find a way*
> *Bring me back to sweet Virginia someday*
> — Virginia Lament, *words and music by Michelle Beall*

"Our identity is tied to our sailboat and coastal marinas, and we return there as much as possible," said Rich, "but we will continue to take gigs across Virginia from Hampton Roads to Front Royal and beyond. We are adding Robert Gabriel – a stand-up bass player – to our group."

Back at Orlean Market, open mics, live music, and the last Friday jams have all picked back up in 2022 (see Figure 30.3). Owner Kia Kianersi says the last Friday jam is his "favorite night of all."

FIGURE 30.3 Jam at Orlean Market.
Source: © Lori Udall

Established in 1905, Orlean Market and pub is an old rambling Victorian-style country house with five rooms, a pub, market, kitchen, and two outdoor patios. It provides the go-to centerpiece for local farmers, workers, and country residents, to come and enjoy a beer, eat burgers, catch up on local gossip, listen to music, or join in and play at open mic or the jam.

Age range at the jams is from teenagers to folks in their 70s. While we are not yet post- Covid at the time, the quest for music draws people out to Orlean Market.

As I drove up and parked at the Market one Friday, I heard the group already jamming, singing "The Old Home Place," a song about a fellow moving to the city for a gal who then ran off with another man.

I starting singing along as I got my guitar out of my car trunk and walked in.

A final note

These stories about ordinary musicians who love their craft show how they adjusted the best they could to COVID-19 and the post-pandemic world. Their stories suggest that artists are highly independent individuals who pursue their interests even under duress. Non-technical musicians have to adjust to the world of ICT platforms and music apps that can help in songwriting and promoting songs. Younger musicians may have an advantage here as they usually rely on smart phones and computers. It turns out that the schooling and other experiences in early life have provided the skills and values needed to endure.

Finally, what was described about musicians coping with COVID-19 life in small town Virginia apply to the many musicians in towns and cities around the world. In poor countries where musicians must rely on their performances to provide daily survival, the stress and pressure to find ways to continue performing and creating new musical forms and songs is overpowering. In these cases, the musicians band together to reinforce self-esteem and provide assistance and hope to get past the societal breakdown due to quarantine isolation. Technology helps musicians survive and in so doing, keeps cultures alive.

31
EMOTIONAL AND PHYSICAL ISOLATION IN A LATINO COMMUNITY

Alfred Anduze

Introduction

The act of socializing for the people who live on the island of Puerto Rico, a US territory in the Caribbean, as in many Latino communities, is the national sport. One might expect major disruptions to social contact and relations to have serious effects on the emotional and physical health of the population. This chapter describes how the island of three million inhabitants living in urban and rural communities survived the COVID-19 pandemic from January 1, 2020, to August 31, 2022. Clearly, social isolation, comprising reduced encounters, interactions, companionship with other people, family and friends, a reduction in direct access to travel, independent movements, communication and learning, led to disconnection, loss of social skills and a sense of community, long associated with stress-related diseases. Luckily, a functional public health system, in place following two disastrous hurricanes (Irma and Maria) in 2017, provided individuals, families, and neighbors with up-to-date medical and psychological care. The island became a leader in percentage of the population vaccinated, low incidence of cases, and low fatality rates.

The etiology of emotion

As a medical doctor, I am aware that a significant reduction in physical and mental stimulation by the gentle human touch leads to anxiety and distress, which may result in loneliness and depression. *Homo sapiens* developed receptors for touch that elicited the release of good-feeling hormones. The initial infant-mother relationship is based on the release of oxytocin (cuddle hormone) from the hypothalamus, serotonin (happiness) from the gut, and dopamine (mood)

DOI: 10.4324/9781032690278-36
This Chapter has been made available Under a CC-BY-NC-ND license.

from the basal ganglia of the brain. It stimulates the growth and development of the neurological system in children and adolescents. Positive social etiquette emerged in late Paleolithic hunter-gatherers with cooperation, altruism, and social skills essential to group survival. The Neolithic settlers that followed were more possessive and created social stratification with its fear of others, misperceived threat, exaggerated hostilities, and open aggression.

These traits often resulted in antisocial personalities, lack of empathy and compassion, selfishness (I remember there was a shortage of toilet paper in supermarkets), and concomitant physical infirmities. Then too, feral children, also known as "wolf children," who grow up without human contact, are devoid of personality, cognitive, and communication skills, and are averse to physical touch. Physical interactions with others are necessary for growth and development, and are extremely effective stress reducers. Note that home-schooled children, without the input of professional teachers and physical and emotional interaction with peers and extended adults, were deprived of the opportunity to hone their social skills. As a physician seeing many children, I noted a distinct reduction in their basic manners, even in the presence of their parents. Without the opportunity to interact freely, their conversational abilities needed much training.

Among modern humans, contact deprivation results in the release of excess stress-related hormones (cortisol, noradrenalin, and adrenalin) with detrimental effects on all tissue and organ systems, especially immune, cardiovascular, cognitive, central nervous, and digestive. The World Health Organization reported 50 percent higher rates of depression, vascular dementia, and Alzheimer's; 30 percent risk of heart disease, hypertension, and arrhythmias; 32 percent higher risk of stroke; and 30 percent risk of digestive issues, gastritis, irritable bowel syndrome, ulcerative colitis, and enteritis; neurodegenerative and autoimmune diseases like lupus erythematosus, amyotrophic lateral sclerosis (ALS), multiple sclerosis (MS), type 2 diabetes, thyroiditis; and cancers associated with loneliness and depression.

The chronic release of cortisol further impairs the immune system by weakening T-cell and interleukin-1 (IL-1) responses and function, releasing overactive cytokines that disrupt signaling between cells, and reducing B-cell function. This results in loss of antibody production, breakdown in normal tissue function, and a further reduction in the reaction to the COVID-19 virus. Loneliness hurts like physical pain, as some neurotransmitters and receptors are the same.

The power of a hug to combat loneliness

The power of an affectionate hug (*un abrazo*) in the Latino community is a traditional necessity, as is language and cuisine. Holding hands, a handshake, a pat on the back, and touching the arm or leg of someone while conversing are well-established traditions within the Latin and Caribbean cultures. Any loss of the opportunity to touch has a detrimental effect on the psyche and physiology of the individual. Hugging boosts the immune system to increase T-cell production, reduces the release of stress hormones, increases feel-good hormones, and serves to reinforce and cement relationships throughout life. Emotionally, it is gratifying to look forward to greeting and leaving a friend and relative, and sometimes a stranger.

Loneliness

Loneliness is the number one global disability and cause of mental and physical decline, often due to lack of human contact and hugs. Lack of friends, companionship, or confidantes may lead to depression that promotes homesickness, sadness, and grieving. This releases inflammatory cytokines that can be just as lethal as obesity and cigarette smoking. Living alone may be acceptable and functional, but feeling alone is strongly associated with disease complexes. Pandemic-fueled under- or overeating, sunlight-deficient sedentary lifestyle, and sleep-deprived social isolation were reported more frequently in young adults and middle-agers than in seniors. Islands and rural areas with less access and travel opportunities reported higher incidences of symptoms associated with contact deprivation (Cacioppo et al., 2011).

The pre-pandemic cross-national studies of loneliness in Germany, Mexico, South Korea, and Scandinavia were better than in the United States, the United Kingdom, Australia, and Japan (18–22 percent of seniors self-reporting), as the former have higher rates of social interactions, more caring for each other, and better government safety nets. Levels of anxiety, depression, and COVID-19 associated morbidities were higher in the lower socioeconomic, underdeveloped countries, and LGBTQ groups where the stress of daily living was obvious (Anduze, 2021). The Latino community in Puerto Rico suffered initially from the loss of physical contact and reduction in mental stimulation until we established more reliable digital communications and grew accustomed to the change. It helped that in a relatively small area, we could still see and hear each other.

As in similar communities, strong cross-generational connections in Puerto Rico held firm and limited loneliness, especially for the elderly. There were 44 nursing homes on the island for three million people. New York State has 90,000 nursing care facilities for 20 million people. With life expectancy in both locales being the same (age 80), the math suggests that the island should have over 400 nursing homes. In keeping with tradition, Latino seniors remain within and receive care from the family unit, no matter their medical status. Another unexpected feature of the Puerto Rican group was that by March 2021, 90 percent of the population had received at least one COVID-19 vaccination, while masking in public spaces was almost at 95 percent (except for the tourists from the mainland), and almost no one remained "alone."

The CFR (case fatality ratio, number of deaths from the disease related to number of cases) were higher in some larger, more developed countries than in smaller, less developed ones. Despite the poverty rate (43%) in Puerto Rico being higher than in the United States (14.7%) in 2021, the island population had a slightly lower CFR. The prevalence of chronic illnesses, obesity, cardiovascular disease, and type 2 diabetes is very high on the island, and accounted for co-morbidities in COVID-19 deaths, as in other countries.

With the increase in world incidence of anxiety and depression (to 25%), and the United States with reports of 59%, Puerto Rico remained at about the same level as the pre-pandemic (20%). We attribute this to the strong family relationships and extensive friendships that kept people grounded and in functional contact with each other (The New York Times, 2023).

Responses to COVID-19 across generations in Puerto Rico

Generation Z

Among Generation Z (6 to 24 years old), 70 percent reported anxiety and depression early on, before diverting their attentions onto their electronic devices, giving up on social skills, and settling into their comfortable world of indifference. Their anxiety indices reduced to 30 percent by the following year. The tendency to "coddle" children even into their teen years, instead of letting them explore and expand their range of interactions, was just as detrimental as not "cuddling." Overprotective parenting may have contributed to the rise in antisocial behavior seen during the pandemic.

Many 13- to 18-year-olds in the United States and Puerto Rico turned to the use and abuse of antidepressants, by both prescriptions and illicit acquisition. Vaping and alcohol use increased in the 18- to 24-year-olds. With schools closed, this group suffered from a lack of opportunity to establish and practice face-to-face social skills and gather valuable information from a variety of people, especially their grandparents and extended family adults. One cannot build relationships based on texting or even Zoom.

As physicians, many of us found that telemedicine, without touch and environmental context, seriously limited our capacity to evaluate and diagnose, treat, and monitor the outcomes of our patients. Sometimes, lack of physical examinations can have permanent deleterious effects on a diagnosis and outcome. University students suffered from a lack of intellectual contact with professors and classmates in places like Peru and India.[1]

Millennials (Generation Y)

The Millennials (or Generation Y), ages 25–44, appeared to ignore the rules of social isolation and went to bars (when open), concerts (when staged), and parties (any time). They increased their use of legal and illegal drugs, resisted masking and vaccinations, and did as they pleased. Their anxiety and depression levels were low in comparison with other age groups, though the incidences of mental and physical illnesses were similar. They handled their insecurities between themselves and supplemented them with mind-altering chemicals and online virtual encounters, in which reality and fantasy interacted.

Generation X

Generation X, ages 45–65, had the highest levels of anxiety and depression, abuse and use of multiple medications, failures of self-care, loss of a sense of community, loneliness, and loss of financial reliance (increased debt). Torn between two generational worlds, this group faced the pressures of taking care of elderly parents and school-age children at the same time. Some young adults of the previous generation moved back into the home, refusing to provide care or take responsibility for loan debts. The middle-aged workforce faced reduced opportunities for jobs and housing, and incurred the highest incidence and prevalence of psychological distress and poor well-being – more in 2020 than in 1990 and more in the United States (Puerto Rico included) than in other countries.

They experienced high levels of inequalities between genders (women leaving work to render family care or being laid off first), and less access to healthcare linked to socioeconomic and ethnicity factors. Under the US system of healthcare tied to employment, many lost their health insurance during the COVID-19 pandemic. Stimulus checks arrived and allowed for survival, nothing else. These "middle-agers," the backbones of their communities, did not have specific programs in place to sustain them.

The entire Gen X population endured the stresses and strains of repeated financial crises, with poverty rates affecting over 45 percent of all respondents. Two major hurricanes (Irma and Maria in 2017) and a series of earthquakes all but shattered their defenses. They searched and scrounged, made things work, and made ends meet to provide basics, but had very little left for themselves. Suffering from the daily stresses that led to poor physical and emotional health, this group had the highest rates of obesity and type 2 diabetes, which contributed to the morbidity and complications of COVID-19.

Baby boomers

Seniors, the so-called baby boomers, did better overall than the Gen X "middle-agers," as they had more experience with catastrophes and more solid social skills to maintain contacts and interactions with family, friends, and acquaintances. Despite being more isolated, often living alone, relatively indigent, with higher prevalence of chronic illnesses, more dependent on others for healthcare, and higher numbers of deaths from all causes, this group kept their chins held high. With their social circles intact, seniors set about the business of adapting to quarantine lockdown and did everything possible to maintain contact.

They observed and learned to use digital technology to maintain old social connections and establish new ones, to amass and pass on information, to entertain and to create.[2] Many seniors ignored the old dog/new trick stereotypes and not only learned but excelled in virtual socializing, group business meetings, medical office visits, ordering and traveling online, and social media networking.

There was a minority antisocial section of society that showed little concern for the dead and dying around them. They preferred to guard their possessions and adopt aggressive stances toward their elderly neighbors, who were already in poor health, experiencing accelerated aging with increases in chronic degenerative diseases, cognitive decline, and premature death. Dying alone was the worst potential outcome for seniors, but members of this group did not seem to care. Acceptance of this attitude seemed to spread along with the pandemic. Not only did we lose our grandparents, some of the glamor and honor that had accompanied the position disappeared as well.

Being a senior

Equally discomforting (to me as a senior) was confronting and accepting the fact that some of the younger generation with the attitude that "old folks are going to die soon in this horrible world that they made anyway" attributed so little value to the lives of the elderly. (In a few weeks in January 2020, I went from being a minority on the rolls of the federal government, with some value as a physician, to the ranks of the expendable retirees on the world stage. It was a dubious transformation, and it hurt.) As many of the deceased were fellow seniors, those left behind became adept at the art of grieving with all its due respect and dignity attached. With more disappointment than anger, seniors grew up faster, became wiser, and were none the worse for it.

Amid the gloom and doom of the early pandemic, there were significant positive gains in awareness of the benefits of the six strategies of health: (i) quality nutrition, (ii) regular exercise, (iii) mental stimulation, (iv) stress control, (v) good habits and avoidance of bad ones, all involving solid (vi) social connections – and that attaining good health increased the chances of a positive outcome from the viral infection and the treatments. Immunity and antibodies became household words. Families and couples grew closer, became more tolerant and understanding, learned to cope and manage together as necessity called. Among the socially aware, there was an increase in empathy, compassion, patience, and gratitude with a discernible appreciation of life and what the five senses provided.

Summing up

We learned from the pandemic that the first requirement for surviving a bad social life situation is a strong physical and mental well-being, and that face-to-face communication and touching are significant factors in growth, development, and maturation. The second requirement is the ability and impetus to manage situations. We learned that using basic social etiquette and acknowledging the existence of others can be a positive health factor. Practicing social skills is essential to developing a full and effective interactive life. Working together to solve problems is far better than doing it alone, and good health can prevent stress-related diseases and increase chances of surviving a deadly viral pandemic. People should learn to use technology wisely and support science-based evidence.

The individual with a solid family and functional stress-free life had better outcomes than the one who lived under duress and alone. The third requirement is to cope with and adjust to those things one cannot control, noting that even if our lives do not return exactly as before, everything will be all right. As COVID-19 variants adapt and proliferate, we know that life in the future cannot be lived on social networks and video calls.

Telemedicine

For me, telemedicine without the element of touch will never be a sufficient substitute for in-person medical examinations. With the resume of office visits, doctor-patient direct contact remains minimal, usually with the computer-generated diagnostic testing in between. Even with the easing of restrictions on socializing, the persistence of contagious variants and steady numbers of new cases and deaths continue to stymie the once vibrant social life of Puerto Rico.

People are more cautious of each other, children still do not know how to say a "hello" in a simple encounter, food deliveries to private homes continue to rise, celebratory gatherings at weddings and funerals continue to fall, and the prolific "hug" is sparse and suspect. We are a social species and may become depressed and seriously ill when lacking secure companionship with physical interactions.

We must unite behind the science and accept the vaccinations for the good of the whole of humanity. (As Puerto Rico was 80 percent fully vaccinated as of March 2022, the number of new cases is holding steady and the case fatality rate continues to fall.) If we remain masked and, where appropriate, choose our gatherings carefully, maintain sensible hygiene, and interact with like-minded people who also practice good public health, then we can feel comfortable.

And guess what! We'll be free to hug again.

Notes

1 See Chapter 7, "Adapting to virtual education during the COVID-19 pandemic in Peru" by Victor Saco and Chapter 8, "The plight of virtual education in India due to COVID-19" by Anwesha Pal.
2 I wrote and published a textbook on this subject during the pandemic. See Anduze, A. (2021). *Social connections and your health*. Yorkshire Publishers.

Reference list

Anduze, A. (2021). *Social connections and your health*. Yorkshire Publishers.

Cacioppo, J. T., Hawkley, L. C., Norman, G. J., & Berntson, G. G. (2011). Social isolation. *Annals of the New York Academy of Sciences, 1231*(1), 17–22. https://doi.org/10.1111/j.1749-6632.2011.06028.x

The New York Times. (2023). *Tracking Coronavirus in Puerto Rico: Latest map and case count*. www.nytimes.com/interactive/2021/us/puerto-rico-covid-cases.html

32
BRAVING COVID-19 THROUGH THE GROSS NATIONAL HAPPINESS WAY IN BHUTAN

Tshering Cigay Dorji

Introduction

Bhutan is a mountainous country in the Eastern Himalayas with an area of 38,394 square kilometers located between China and India. It has a population of about 750,000 people, most of whom are Buddhists. It is guided by the principle of Gross National Happiness (GNH) and is enshrined in the Constitution of Bhutan. GNH ensures that the government follows a more holistic approach to development, balancing material progress with the spiritual well-being of the people.

As a small country with limited resources, Bhutan could have reeled under the pressure of COVID-19, the most challenging crisis the world faced since World War II, as the UN Secretary-General Antonio Guterres termed it. However, thanks to the compassionate leadership of our king, His Majesty Jigme Khesar Namgyel Wangchuck, who architected the whole COVID-19 response strategy in the spirit of GNH, and led from the frontlines by frequently visiting the red zone areas to comfort the people and motivate the workers, Bhutan was able to brave the pandemic much better than many other countries.

The COVID-19 pandemic brought the world and communities together and, in many instances, brought the best of each nation and every individual. There have been many heartening stories of love, courage, and sacrifice from around the world. Bhutan has her own fair share of such stories to share.

Compassion and clarity of purpose

Compassionate leadership and clarity of purpose echo as the overriding theme in Bhutan's fight against COVID-19 since it first surfaced. From the beginning, Bhutan was clear that the health and safety of the people should be

DOI: 10.4324/9781032690278-37
This Chapter has been made available Under a CC-BY-NC-ND license.

accorded the greatest priority. "We will have the opportunity to rebuild the economy, but not life, once lost" was its guiding mantra.

Bhutan detected its first case of COVID-19 on March 5, 2020, and was confirmed the next day. The patient was a 76-year-old tourist from the United States, who had entered the country from India on March 2.

Contact tracings were launched per the patient's itinerary. It included people he associated with at all the points and stops along the way. The close contacts were his partner, the driver, and the guide. All three were asymptomatic, but they were quarantined immediately. Only his partner tested positive ultimately.

National Preparedness and Response Plan

Per the National Preparedness and Response Plan, the country stepped into the "orange" zone, according to which the government has to isolate confirmed cases, quarantine suspects, and carry out closure of schools, institutions, and public gatherings in the affected localities. The government closed schools and institutes in Thimphu, Paro, and Punakha for two weeks with effect from March 6. It also imposed two weeks' restriction on all incoming tourists with immediate effect, but this restriction remained on indefinitely as the global pandemic situation worsened in the days to come.

The COVID-19 patient was given the best of care. As his condition improved, he chose to go back home by a chartered flight. In a video message sent in after he recovered fully, he acknowledged how the king personally ensured that he was cared for and given proper medical treatment. Later, the patient sent a touching video message (Kuensel, 2021) in which he recalled how his life was saved with timely and quality care and treatment (Tshedup, 2021).

On March 22, 2020, His Majesty gave his first royal address to the people of Bhutan since the start of the pandemic. He announced that the border of Bhutan would start to close the next day for our safety because of the worsening pandemic situation globally and in the neighboring region:

> At such a time, the health and safety of the people of Bhutan is of the greatest priority, and as such, we are putting in place every measure necessary to safeguard the people of Bhutan. Should those of you who are abroad at this time wish to return home, the government will help you. I ask those of you who are studying or working abroad, not to worry.
>
> COVID-19 will cause great disruptions to the global economy, and Bhutan will not be an exception. The economic repercussions will not just impact a select few sectors, but each and every one of us. At such a time, we must exhibit the strength that comes out of our smallness, remain united, and support one another. During such exceptional circumstances,

the government will take the responsibility of alleviating any suffering to the people due to the virus.

(The Bhutanese, 2020, paras. 4–5)

In his second national address on September 12, 2020, which was right after the first lockdown Bhutan underwent in August 2020, His Majesty reiterated, "During this pandemic, my only priority is the well-being and happiness of our people, including those living abroad" (Druk Gyalpo's Relief Kidu, 2020, para 17). True to these words, life and livelihood of the people were accorded the top priority throughout the pandemic situation.

Livelihood support

As the pandemic caused immense social and economic difficulties and tragic loss of lives across the world, Bhutan was acutely aware of the economic difficulties our own people would face.

The National Resilience Fund was set up in April 2020 to provide economic relief to people through the Druk Gyalpo's Relief Kidu. Kidu is a Bhutanese term that stands for "welfare." The Relief Kidu granted monthly income support to individuals and loan interest payment support to borrowers for a period of one year (April 2020–March 2021).

Interest payment support (IPS) provided across-the-board relief to all borrowers, including non-performing loan (NPL) clients. NPL clients include borrowers who have not been able to pay their monthly EMIs for three months or more. A hundred percent IPS was granted for the first six months followed by financial service providers, providing 50 percent of the IPS. Additional monetary measures, such as deferment of loan repayments and provision of working capital and short-term loans at concessional interest rates, were rolled out to ensure that relief measures were adequate and inclusive.

Within that one year, the support had helped about 52,644 individuals sustain livelihoods and more than 139,096 loan account holders from the interest payment support.

Of the 139,096 loan account holders that benefited from the support, 17,766 were NPL accounts. The total number of beneficiaries includes 60,002 salaried individuals and 79,094 business accounts. Of the BTN 11.064 billion granted as kidu from the interest waiver facility, BTN 1.15 billion was granted to salaried individuals and BTN 9.91 billion to businesses.

In April 2021, the Druk Gyalpo's Relief Kidu (DGRK) was extended for another 15 months (from April 2021 to June 2022), in view of ongoing social and economic difficulties faced by the people due to the COVID-19 pandemic and the uncertainties stemming from regional and global developments (Wangchuk, 2021).

Compassion in action

Bhutan has many street dogs. Since we, as Buddhists, refrain from killing, controlling the dog population has been a challenge. These dogs survive on the kindness of the people who hand out leftover food. During the first nationwide lockdown imposed on August 11, 2020, many people's hearts went out to them, wondering how they would survive without people being able to venture to feed them.

But this had already been thought about. The Royal Bhutan Army began a nationwide stray-dog feeding program to ensure that dogs do not starve during the lockdown (see Figure 32.1).

As instances of domestic violence increased during lockdown, the government set up shelters for the needy or victims or those at risks of domestic violence to come and stay during the lockdown.

Desuung – Guardians of Peace

Desuung, which literally means Guardians of Peace in Dzongkha, the national language of Bhutan, is a one-month integrated training program instituted by His Majesty. It is built upon the spirit of volunteerism and the positive influence of ethics and values of community service, integrity, and civic responsibility. The trainees are known as *Desuups* once they complete the training.

FIGURE 32.1 A *Desuup* volunteer feeds the dogs in the central district of Bumthang during the lockdown on January 18, 2022.

Source: (Picture courtesy: Bhutan Broadcasting Service Facebook Page)

A *Desuup* is expected to actively volunteer during disaster operations, participate in charitable activities, and be of service to others throughout their lives.

During the pandemic, hundreds of people, especially the youth, became *Desuups* and came forward to volunteer and serve as frontliners. As of August 31, 2022, the *Desuung* website (https://desuung.org.bt) showed the total number of 30,311 *Desuups* (18,575 male and 11,736 female).

The *Desuups* played a crucial role in the fight against the pandemic. They manned the quarantine facilities as well as the border posts in the south along the long and porous border with India. They also manned the call centers, guarded the quarantine and isolation facilities, helped distribute food and medicine during the lockdowns, guided the movement of people, ensured that the Covid protocols were followed, and helped in the mass vaccination campaign.

Desuung Skilling Program

His Majesty was deeply appreciative of the services rendered by the *Desuups* during the pandemic and wanted to provide them something useful in return. Many of the *Desuups* were young people who previously worked in the tourism sector and were now unemployed, while others had finished college or school.

The *Desuung* Skilling Program (DSP), a massive national skilling program for the *Desuups* in various fields, was instituted in 2021 to provide skills-based training to the *Desuups* enabling them to earn better livelihood.

As of August 31, 2022, DSP has trained 3,679 *Desuups* (Male – 2,051, Female – 1,506) in various skills, ranging from arts and crafts, culinary arts and carpentry, to robotics and full stack software development.

The power of prayer

Bhutan is a deeply spiritual place. There is a central monastic body called the Zhung Dratshang supported by the state. The Zhung Dratshang has presence in every district. In addition to the monks of the Zhung Dratshang and its monasteries, there are many privately owned or community-owned monasteries, temples, monks, and Lamas who are not part of the Zhung Dratshang. During the pandemic, these religious communities came together to perform rituals for the well-being of the people.

For the Bhutanese, these rituals provided much needed psychological comfort. The Bhutanese believe everything that happens is a result of the interplay of myriad causes and conditions. Many of these causes and conditions are beyond our control. The prayers and rituals are expected to help align the causes and conditions in a right manner to bring about the desired result.

It is with this belief that the day for the start of the first vaccination drive was chosen after careful consultation with the astrologers. They chose March 27, 2021. On top of that, the first person to get the vaccination, a person of particular age and gender, was also chosen based on the advice of astrologers.

Highly successful vaccination drives

In the first round of vaccination drive, Bhutan vaccinated 478,829 people out of the total 496,044 eligible population aged 18 years and above, across 1,217 vaccination centers in three weeks, achieving a vaccination coverage of 96.5 percent (Tsheten et al., 2022). The second round of vaccination kicked off on July 20, 2021, again on an auspicious day per Bhutanese astrology. A cumulative total of 473,715 people were vaccinated within two weeks, covering 95.6 percent of the eligible adult population.

Bhutan managed to inoculate more than 95 percent of its eligible populations in two rounds of vaccination campaign. According to Tsheten et al. (2022), enabling factors of this successful vaccination campaign were strong national leadership, a well-coordinated national preparedness plan, and high acceptability of vaccine due to effective mass communication and social engagement led by religious figures, volunteers, and local leaders.

Bhutan had reported a total of 2,596 COVID-19 cases and only three deaths as of September 15, 2021. Many countries and international organizations praised Bhutan for its remarkable achievement in containing the pandemic (Ongmo & Parikh, 2020; UNICEF, 2021; WHO, 2022).

Technology to the rescue

On the morning of March 18, 2020, H. E. Lyonpo Dechen Wangmo, the minister of health, convened a meeting with a few IT professionals (including the author of this paper) at her office. Her message was clear and simple: "What can the IT fraternity do to help the government combat COVID-19?"

The ministry needed tools to collect and process information related to border crossings, people visiting the flu clinics and quarantined people, and a GIS-based dashboard for the decision-makers.

In response, the IT professionals from government agencies as well the private sector came together in the true spirit of collaboration and patriotism, and developed some of the most useful software applications within a month or two (see Figure 32.2). The first successful application rolled out was Druktrace, a contact tracing app built by Bhutanese programmers without any outside help. Others followed, which included a quarantine monitoring app, a GIS-based dashboard to view the COVID-19 status in all of Bhutan at a glance, a system to register all inter-district travelers, and a registration system for vaccination.

FIGURE 32.2 IT Teams busy at work on the night of March 19, 2020, Bhutan.
Source: © Tshering Cigay Dorji

The most heartening thing in this whole initiative was the fact that the people were only too happy and willing to come forward to help. Even those from the private sector came forward without expecting any compensation. Their employers had sent them to join the initiative, leaving behind other work in their offices. This is an example of the unity and solidarity shown by people from all walks of life during the pandemic. We even had cases of farmers donating rice and vegetables to the frontliners.

Learning and the way forward

As every adversity offers a lot of opportunities to learn and grow, the COVID-19 pandemic brought out the best in each nation, and every one of us as individuals. In Bhutan, we saw increased sense of cooperation among the different sectors, agencies, and individuals. There were some very good examples of collaboration between the private sector and public sector. There was also a heightened sense of unity and patriotism and heartfelt gratitude to His Majesty, who put his own health at risk by traveling numerous times to the red zone areas to comfort the affected people and inspire and encourage those on frontline duty.

During the pandemic, the bureaucracy demonstrated willingness to experiment and innovate, listen to the public, and work collaboratively with the private sector. We should not drop these good practices once the pandemic is over.

The pandemic also brought to the fore our weaknesses, such as our overdependence on imports, our weak economy, and lack of food self-sufficiency. We have a long way to go as a nation to build a stronger economy and improve our food security. With this realization, Bhutan has initiated major reforms and transformation exercises, which are currently under way.

Reference list

The Bhutanese. (2020, March 28). *His majesty's address on 22 March 2020 announcing the closing of Bhutan's borders.* https://thebhutanese.bt/his-majestys-address-on-22-march-2020-announcing-the-closing-of-bhutans-borders/

Druk Gyalpo's Relief Kidu. (2020, September 14). *His majesty the King's address to the nation.* https://royalkidu.bt/his-majesty-the-kings-address-to-the-nation-12-september-2020/

Kuensel. (2021, February 21). *A king to a tourist's rescue* [Video]. www.youtube.com/watch?v=0_2qCgv6m04

Ongmo, S., & Parikh, T. (2020). What explains Bhutan's success battling COVID-19? *The Diplomat.* https://thediplomat.com/2020/05/what-explains-bhutans-success-battling-covid-19/

Tshedup, Y. (2021, February 21). A king to a tourist's rescue. *Kuensel.* https://kuenselonline.com/a-king-to-a-tourists-rescue/

Tsheten, T., Tenzin, P., Clements, A. C. A., Gray, D. J., Ugyel, L., & Wangdi, K. (2022). The COVID-19 vaccination campaign in Bhutan: Strategy and enablers. *Infectious Diseases of Poverty, 11*(6). https://doi.org/10.1186/s40249-021-00929-x

UNICEF. (2021, May 22). *Bhutan's success against COVID-19 desperately needs preserving* [Press release]. www.unicef.org/bhutan/press-releases/bhutans-success-against-covid-19-desperately-needs-preserving

Wangchuk, R. (2021, April 4). Royal Kidu benefited 52,644 individuals and 139,096 loan account holders. *Kuensel.* https://kuenselonline.com/royal-kidu-benefited-52644-individuals-and-139096-loan-account-holders/

World Health Organization, Regional Office for South-East Asia. (2022). *The people's pandemic – how the Himalayan kingdom of Bhutan staged a world-class response to COVID-19, 2022.* World Health Organization. Regional Office for South-East Asia. https://apps.who.int/iris/handle/10665/362219

33
A RESURRECTION
Human connections and beyond

Suresh Nanwani

My many forms in the pandemic

The pandemic has altered many ideas for me, and I guess for others too. For me, in simple words, it was a nightmare, a Freddy Krueger. However, at the end of the spectrum of the mysterious and dark night, it is the glowing sunlight at dawn, which I wake up to see and appreciate, or an arched rainbow with psychedelic colors splashed across the clear blue sky or on the odd day, a murky grey sky. These variations are treats for me in the past two and a half years. From January 2020 to June 2022, at the time of my writing, I have reinvented myself. The world around me seemed to collapse – the people I knew, my friends and colleagues. I was cocooned, trapped, and confined in my own world (this sentence reflected my sentiment in the early stages of COVID-19). After a year of COVID-19, I rephrased this sentence as I was cocooned and trapped within, but from within, I reoriented myself to unleash positivism – I connected or reconnected with society and beyond.

My journey over the past few years has seen a different context of me and my surroundings. It was a rude shock to hear that my PhD graduation ceremony was cancelled because of the lockdown in March 2020 (the call-off was announced two days before the lockdown was imposed). I thought at that time the lockdown was temporary, but it stretched to over two years. I have not been out of my unit beyond a 3-mile radius for the past two and half years, and any distance beyond a half-mile was for essentials like medical or dental treatment.

My condo unit was my home, my refuge, my everything. And within that unit, I had three friends I connected with – my computer, my iPhone, and my

DOI: 10.4324/9781032690278-38
This Chapter has been made available Under a CC-BY-NC-ND license.

printer. They were the best of company, and through them, I connected with my human friends and the world. I also wrote two books about my transformational life journey (Nanwani, 2021) and human connections in teaching experiences (Nanwani, 2022).

The first two years of Covid

The first two years were filled with uncertainty, ignorance, and helplessness. **Uncertainty** as many things were put on hold, such as normal daily activities. **Ignorance** abounded as I did not know what was really going on with the pandemic, and I accepted wearing a mask as this was in line with public health requirements by the government and health authorities. **Helplessness** as I still remember with horror when I visited the ubiquitous 7/11 convenience store to buy some essentials in the early days of the lockdown – the cashier coughed and I realized she did not have her mask on. She apologized profusely to me and my friend (there was no one else in the store), saying the straps of her face mask had broken and she had no backup mask. I gave her a spare mask I had on me, and she thanked me profusely.

That encounter pained me and gave rise to many thoughts. She had to work for a living, attend to the needs of customers in the convenience store, and manage with or without a mask (her boss did not provide her with a mask, she said). On the other hand, I was fortunate to have a mask, and as I was retired, I did not have to work. She, on the other hand, had to work for herself and her family. There was no equilibrium.

As I pondered on this, I was grateful I did not have to work as I had retired a few years earlier, and am now working by writing and teaching. But there were others around me who had to work. There was Manuel, who lived in the slums and had to work for a living to help his ailing mother and sibling who was recovering from a kidney transplant. He worked as an IT support in a local bank and could ill-afford to miss his work. I loaned him money so he could buy a scooter to commute to work (the public transport of mass rail transit and buses were not dependable and safe). There was a female security guard, Lenie, from my condo, who delivered her baby boy (named Covid) in March 2020 and then carried on working to feed her family. These awakenings made me realize that I had so much to be grateful for – I was 65 and Manuel and Lenie were in their 30s – as I did not have to work and contend with difficulties in going to work and supporting a household.

My realization: I am in a cocoon

By living with myself, I realized I was in a cocoon. And withdrawing myself and staying within this encasing did me no good at all – I was trapped with uncertainties and needed to free myself. That freeing experience was a wake-up call: I had to tell myself I was alive and kicking, and convince myself that I could creep out of the deep hole I had fallen in. Fear was my worst enemy. I looked around . . . my computer friend connected me with people I have known from 1981 to 2003 but then lost contact with. It was surreal, to be reconnected to Owen from the Bahamas after a 20-year gap and Samuel from Perth after a 40-year gap.

Reconnecting with Owen was one of the "best" things the pandemic gave me, as I think I would otherwise have not reconnected with him. With renewed connection, I got to know his daughter Meghan, whom I had met when she was seven, and is now in her late 20s. She married in the Bahamas as her father Owen beamed with joy in the video and photographs taken during the wedding, which had few attendees due to the pandemic restrictions. Reconnecting with Samuel was a blessing – IT technology helped and we located each other through the internet, as there was a long absence of communication while we pursued our careers and somehow lost connection. Now there was reconnection and renewed joy (and it took a pandemic to bridge and rekindle the friendship!).

Reflection of things past

I had time to reflect and in that process, unearthed thoughts that were deeply recessed in my consciousness. I revisited my viewing of Shakespeare's *King Lear* (written about 1606). The play had references to the bubonic plague – Goneril, one of Lear's three daughters, was "a disease" in him, a "plague-sore." That reference never struck me till now, though I read the play 50 years ago. Then there was Shakespeare's *Romeo and Juliet* (written about 1595), and it dawned on me now about the references to the quarantine. Juliet confides in Friar Lawrence, and he finds out that the messenger he sent to Mantua with a letter to Romeo explaining that Juliet is alive has been quarantined because of an outbreak of the plague. Had it not been for the quarantine, the ending for the star-crossed lovers would well have been different. But my mind shifts 400 years forward to the present.

How did I get to gloss over these two matters? Goneril being a "plague-sore," not just an unfilial and treacherous daughter. And the lives of Romeo and Juliet would have been totally different if there had been no quarantine, and we would have missed the story of these immortalized star-crossed lovers. The term "quarantine" was initially used for ships arriving at Venice, which had to anchor for 40 days before landing. The oversight was simple – a lack of awareness hid it from me, and it took the pandemic to make me revisit my beloved plays with renewed eyes after 50 years!

Only connect

The pandemic also taught me how to reconnect with long-lost friends. It made me think of two words – only connect. "Only connect" is the narrative of *Howards End* by E. M. Forster, referring to the contrast between two families set in the class warfare of Edwardian England in the 1900s. This refrain is by Margaret Schlegel, from one of the two families. It came to my mind inexorably, churning and tossing my thoughts: "Only connect! That was the whole of her sermon. Only connect the prose and the passion, and both will be exalted, and human love will be seen at its height. Live in fragments no longer." I read this novel in the 1990s, and since then, I have been caught up and inspired by these two words in many contexts. But Covid times perpetuated and sealed these words in me.

Two words, and they make a difference to me – only connect – and I value personal relationships better. Margaret longs for people to communicate with, overcoming boundaries such as class and gender. For me, COVID-19 put a slant on this refrain – how does one connect during the pandemic? Through the telephone, smart phone, computer, with devices and apps like WhatsApp, Messenger, Facebook, WeChat, Line, Skype, Zoom, MS Teams, and TikTok. Before the pandemic, apps like Zoom and MS Teams were hardly used, but now they are here to stay. Isolation abounds within the person, within persons in the same household, among colleagues who work from home or who may be in the office but are muzzled with a face mask.

With these communications, people connect, strangers, family, friends, and colleagues. Connection can be through speaking, dancing, singing, and emojis. I met long-lost friends, acquaintances, new friends, my existing cohort of people – and viewed them virtually. There was connection and personal relationship is placed at the fore if I wish for "real" connection minus holding the hand of the person, minus physical meeting when parties are not in the same country – these barriers are brought down through connecting virtually. A new sense of connection emerges.

I did online teaching during the pandemic in 2020, 2021, and 2022 through webinars and seminars. (While there was a merging of virtual and face-to-face teaching in some universities, I continued with my online teaching as it was not conducive to travel with the travel restrictions.) I found online teaching challenging, where developing emotional skills and classroom feel was critical for the students and for me. Listening to students took a new form – I remember a few students in South Africa and India could not attend some classes due to lack of internet connection or power shortage.

Yet through the internet, there was learning and information flow, meetings and greetings, and cultural understandings. With the pandemic, there was a need to adjust to students' needs who could not submit essays online on time due to power outages or falling ill. I had caught COVID-19 and was grateful I had not caught it during my online teaching sessions.

Personal relationships and meaning

The value of personal relationships is enhanced and cherished as parties enter a new zone and frame where there can be deceptive backgrounds of a beach or a mountain. Personal relationships are tested on new premises – do we *really* connect? What values are placed in connecting? Caring, compassion, resilience, adaptability, empathy, honesty, active listening – there may be many others but I single out the ones I wanted to focus on to better connect. If I just listen actively to my friend, that may be the "best" connection, as she wanted me to listen. The personal connection may also be most effective in just seeing each other virtually. Over the pandemic period, new forms of connectivity arose, such as webinars or live conferences or hybrid meetings. All these point to safety and the comfort zone of the participants.

Ikigai

Attendant to connecting with humans and valuing personal relationships, I found connecting with nature and the environment around me take shape in a new form I had not imagined or realized before. When I saw birds instead of airplanes, it made me feel that the birds were resuming their "right" to fly without intrusion from manmade machines. It was a wondrous sight to behold a flock of birds swirling in the sky. I wondered, is there a Jonathan Livingstone Seagull (just like in the book of the same title by Richard Bach) in me that seeks perfection in leading a peaceful and happy life? I asked because I was pondering over many questions.

Do the birds know that there is COVID-19 and that is why there are less flights and more space for them to fly unrestrained by interference created by humankind? Do the trees lining busy streets in the city feel different now, as they have less pollution to contend with due to less vehicles plying in commercial areas? Are the fishes in the pond in my condominium hungry, wondering why there are less people feeding them during the early pandemic days?

Ikigai ("iki" is to give, and "gai" is to reason) is a Japanese term, a mindset that blends appreciative inquiry with flow and creativity. Ikigai resides in matters simple or big. COVID-19 brought *ikigai* to me, for me to feast on simple things like sipping hot, freshly brewed coffee and watching birds fly and soar. With a positive mindset, I would imagine myself as a fellow bird feasting my eyes over the land below, even though I was in my unit unable to go out.

That refreshing perspective made me realize how I view the spectrum of things in life and the richness of ideas to appreciate and enjoy life. I learned one salient lesson from *ikigai*: Start with simple, small pleasures like drinking coffee, smell the delicious aroma, be at peace and harmony with my surroundings, and enjoy being in the here and now when engaging in the activity. Feeding the fishes in the pond and watching them speed to feast on food is a beautiful sight to connect with nature and animals.

Myself – before and after Covid

My senses took a different turn. Before the pandemic, it was me, me, and me alone in the pre-Covid world. With the onset of COVID-19, it became me, others, and nature. I feel I have been resurrected from my slumber, and in a new world, I had to reorientate myself and discover new interrelations with society and nature. I do not exist alone when I am one in the scheme of things. My values have changed – I ask myself and reassess my values in life. Health, compassion, and spirituality mean much more to me now than before.

As a Hindu reading about spirituality and the soul, I realize during the pandemic that what I see is transient and ephemeral. I believe that God is eternal (Charan Singh, 1964). Self-realization is crucial to my sustained belief in God. Human life is a journey, not a destination. I learn to connect and cherish for human flourishing.

My priorities have changed with the pandemic. My sense of awareness has reached new heights to find the "me" in myself. Where does this search end? In fact, the search never ends! Answers from my search will lead me to new experiences.

Appreciative inquiry

I practice appreciative inquiry to bring out the best in the person I am communicating with. This has a positive and resounding effect; both the person and I benefit, and we exude positivity at the end of the exercise in discovering ourselves.[1] Conversely, there may be situations when I find someone tells me: Thanks for listening to me. I did not want your inquiry, sympathy, or empathy. I just wanted you to be there to hear me out and say nothing until I let you speak after I have finished pouring out my feelings, my thoughts, and my sentiments. There may be other scenarios.

Beyond her thank you, the lady at the 7/11 convenience store may have had another message for me. She received the mask and thanked me profusely, but what was the real message she wanted to say? I may well have missed something and I could have had an appreciative dialogue or inquiry with her to bring out her feelings and sentiment. That communication would have brought out the best in us – the positivity in her in moving forward, and my knowing that I unleashed my mindset to be receptive to her situation. I need to be grounded more in my resurrection and see the world and nature in different forms so I can understand myself better and relate better with society and nature.

A new day dawns – positive thinking

Having a positive mindset helps – it's stating the obvious but emphasizing it feels good as I need to nourish myself with positivity. In turn, there is a nourishing of humanity when my fellow beings and I connect and we understand ourselves better through values of good social interactions, friendly connections, and additional values we appreciate more during Covid times, such as patience and compassion.[2]

I learn to build on my emotional strength and direct my energy more to humanity. Pre-pandemic, the road was "I was, I am, and I will be." With the pandemic and post-pandemic, the road has become a journey to survive with the refrain, "We were, we are, and we will be." I heard the lady at the convenience store has left. I have lost connection with her, and hope she is well. I smile now when I see birds in the sky flying high and airplanes flying higher – there is resumption.

The encounters, human and nature, in my voyage now are powerful as they make me see the wider perspective of the world I live in. I traverse my voyage with gratitude and compassion, and am learning new traits or values to make me adjust to the new normal. Technology helps me in preparing myself better for the future. I am acutely aware of my need to be more in touch with technology innovations to improve my connections, my teaching, and my learning.[3]

I understand better now when I see myself as part of nature, so when I wake up I smile to see the glorious sun streaking its rays over me. I feel the light is blessing me and also warming me to get up to a bright new day. I know not what will unfold during the day, but that's no matter. There is appreciative living of life, and the day radiates hope and brightness, and even if it rains later, who knows? There may well be a stunning rainbow to appreciate the varied colors and the positive radiance. I need not look for the pot of gold; the rainbow already adds a wondrous touch of nature that I am part of and share the joy of connectivity.

Notes

1 See Chapter 22, "Unlocking from lockdown: reframing the future through appreciative dialogue" by Keith Storace.
2 See Chapter 21, "How a Gen X became a Gen Z at heart" by Bina Patel and Chapter 25, "Self-coaching for pandemic survivors" by Vikram Kapoor.
3 See Chapter 29, "Digital technology during COVID-19 in global living educational theory research" by Jack Whitehead.

Reference list

Charan Singh, M. (1964). *Spiritual discourses.* Radha Soami Satsang Beas.

Nanwani, S. (2021). *Organization and education development: Reflecting and transforming in a self-discovery journey.* Routledge. www.taylorfrancis.com/books/oa-mono/10.4324/9781003166986/organization-education-development-suresh-nanwani

Nanwani, S. (2022). *Human connections: Teaching experiences in Chongqing, China and beyond.* Amazon Kindle. www.amazon.com/dp/B09ZJ5RY36/ref=docs-os-doi_0

PART VI
Conclusion

34
SOCIAL STRUCTURE ADAPTABILITY TO THE PANDEMIC

Suresh Nanwani and William Loxley

Lessons from the pandemic

This book project sought to know if the pandemic made us pause and look at our lives and ourselves more closely (Maslow, 2011). Respite from the daily grind caused by the lockdowns revealed interesting revelations about education and training, health, career and family, public policies and accountability mechanisms, and cultural values. We saw how these events alter human history, including technology, leisure, common values, knowledge flow, demographic shifts, mortality, and more. As a result of COVID-19, (i) new modes of learning skills are replacing the old education system; (ii) governance and policy are reshaping how we fight disasters and treat vulnerable groups; (iii) business and enterprise are revamping work environments to meet generational skills and values; and (iv) ICT is altering the flow of ideas and creativity in redefining cultural thought processes.

In November 2022, WHO worked on a proof of concept for an interactive composite road map that will include various pathogens piloted in 12 countries, which are not yet identified (Ying, 2022). Dr. Maria Van Kerkhove, a WHO official, stated that countries need to leverage the trauma of COVID-19 right now to gear up for the next threat. Otherwise, we would all lose the opportunity the next time a crisis, such as disease X, arises. Five Cs need to be in place to keep us ready for the next crisis: (i) collaborative surveillance to pick up new diseases; (ii) emergency coordination, such as national action plans for prevention preparedness; (iii) access to countermeasures, such as fast-track research and development; (iv) community protection; and (v) safe and scalable care. Two renowned economists, Lawrence Summers and N. K. Singh, have highlighted another pandemic is likely in the next generation. The world must be ready.

DOI: 10.4324/9781032690278-40
This Chapter has been made available Under a CC-BY-SA license

What is the greatest lesson from the pandemic (Charumilind et al., 2022)? People, society, governments, and institutions will have different responses depending on what they experienced with COVID-19. It is not just "being ahead of the curve" but also being in the frame of mind to move forward individually and collectively. This is where mental well-being and coaching, including self-coaching, play a role.[1] For the individual, it is the inner self that needs to relate and adjust, and society and government are instruments fostering individual readiness.[2]

The big question remains as to whether the impact of COVID-19 has altered thinking about public health and medical science warnings. How best can society protect the public without destroying the economy? How well can society assess the damage caused by social isolation on students, research, work, and online socialization during disasters? If society has learned any lesson well, then COVID-19 will have improved the chances that society can survive future disasters where education and knowledge play a valuable role in overcoming adversity.

The central message of this book helps the reader to examine how COVID-19 reshaped societies around the world. Communities, societies, and nations structure themselves around human activities in daily life. Each choice varies for different people by age, gender, residence, educated level, and many more, as they avail of a social structure that generates opportunities and new experiences. This is why the four themes in the book help define key opportunities that structure human activity.

Sector themes affecting institutions and individuals

In earlier chapters, we devised a background framework of the pandemic and presented essays about change to identify relationships that connect people and organizations, which contribute to larger social networks such as whole societies, nations, and humanity. At the heart of social systems, the analytical approach to monitoring society identifies organizations within economies, polities, social communities, and cultural venues, including schools, government, business, and the arts and sciences (Parsons & Smelser, 1956; Wallerstein, 2004; Ryan, 2018; Rutherford, 2019; Haas, 2021; Lupton & Willis, 2021; Ressette-Crake & Buckwalter, 2022).

These institutions reinforce one another by sharing traditions, values, and interests so that society can come together through knowledge and social awareness to solve pressing issues. Institutions within sectors allow themes to encourage adaptability to the environment, set goals, integrate society, and maintain patterns of stability and change, which allow education and technology to play important roles in defining societies.

Structural functionalism suggests that in times of crises, as in pandemics and wars, the main function of institutions is to balance the social structure itself (Parsons, 1977). An alternative approach in sociology is to analyze social change through conflict theory (Collins & Sanderson, 2010), which emphasizes social inequality derived from power and wealth. These factors serve as major societal hurdles to progress when interpreting how parts of society work in ways that lead to conflict across class, gender, ethnicity, etc. Both approaches may be helpful to examine social mobility, governance, labor market, and technological innovation in times of COVID-19.

Based on systems analysis, we settled on four main social processes that have been adversely affected by the pandemic. These processes are likely to have a permanent impact for a generation and alter human activity in ways dissimilar to present-day norms. These major trends focus on (i) educational opportunity causing social mobility; (ii) better risk management of public policy through educating the public; (iii) diversity and inclusivity in career opportunity in work; and (iv) technology for pursuing human development through the internet or social media. What did the chapters in the four themes reveal about how the isolation caused by COVID-19 influences these basic human processes experienced by society?

Opportunity and mobility

In Part II, we noted that the community evokes social mobility as a way students, workers, business people, artists, thinkers, innovators, and policymakers can solve social problems. Higher education institutes in rural India cited in Anwesha Pal's essay present challenges for poor youth seeking degrees in the age of COVID-19 restrictions. Similar issues were raised in the essays by Victor Saco from Peru and Tsotang Tsietsi from Lesotho. The theme of social mobility through educational opportunity is the mechanism that allows the sector to function and society to thrive. Chapters by Frédéric Ysewijn from Belgium and Ariel Segal from Israel tell about difficulties in finding careers after graduation during the pandemic.

The issue of acquiring skills from talent among the young and gaining awareness from knowledge among seniors was raised in William Loxley's essay. The process of acquiring information and knowledge provides skills and expertise that generates more experience, which leads to greater opportunity for mobility. This fundamental process is often unappreciated. Social isolation caused by COVID-19 was bad for education, face-to-face learning, for aspirations, and for most groups, from youth to the elderly. The theme on educational opportunity included six essays: two on career options and remote learning among youth; three on problems faced by higher education institutions in Peru, India, and Lesotho; and one on lifelong learning potential for all age groups, including the elderly who suffered loss of learning during quarantines.

Learning has been challenging for young people at the point in their lives when they choose career and seek employment. Remote learning in the era of adaption to ICT protocols forced many to adapt to new ways of learning.

For those undergoing training and seeking information from the internet, major adjustments to artificial intelligence have resulted in new experiences. For example, the traditional classroom method collapsed and impacted young children the most, as they must acquire socialization skills to proceed further in life. In higher education, mostly rural and poor students struggle to keep up with technology because they lack resources and access to online learning.

COVID-19 has forced instructional methods to change especially through technological inventions. This process has been sped up. The future requires everyone to keep learning to gain new skills and improve their personal wellbeing. The shock of what COVID-19 did by halting skill acquisition led to deficits in expertise. The result is a wake-up call to improve schooling outcomes even as classroom learning becomes more flexible and able to accommodate differing learning styles. Finally, educational opportunity powers lifelong learning that affects the young and old, rich and poor, male and female entering the metaverse.

Public policy and crisis management

In Part III, the reader learned that court systems in Nairobi have gone digital to circumvent collapse of operations caused by COVID-19. Setting policy is the mechanism that allows the polity to function. The set of essays in Part III deliberated on issues related to operationalizing accountability in the distribution of resources. Chapters by Tracy Epps of New Zealand and José Guilherme Moreno Caiado of Brazil dealt with international cooperation themes found in the sociology of international development, including institutional organization and networking. Two chapters also looked at the legal system and legal education issues affected by the pandemic – the first chapter by Leyla Ahmed of Kenya looked at court systems under lockdown, while the second chapter by Shouvik Kumar Guha from India looked into Covid effects of virtual learning on legal education outcomes. Owen Bethel from the Bahamas also reflects on the need for sound public crisis management in his essay.

Li Xudong's chapter looked at the use of big data in monitoring pandemic outcomes in China to explain how the government handled COVID-19. Two chapters looked at top-down approaches to crisis management that create implementation problems when human rights or civil liberties are violated. Antonio G. M. La Viña's chapter assessed the success of COVID-19 mandates in the Philippines, while Nicole Mazurek of Australia described how citizens reacted to the lockdown. Should mandates be forcefully imposed or voluntary? Should face masks be required in public?

Generally, government agencies make better policy choices when they previously have had success in handling disasters such as SARS. COVID-19 erupted suddenly and forced policymakers to make quick decisions on public health, education systems, technological applications, social welfare, and many other processes. These all were influenced by the economic shutdowns, financial debt burdens, and ultimately medical decisions on vaccine distribution to vulnerable groups.

The impact of technological innovation became obvious especially in developing countries where big data did not exist or was totally unconnected to policymaking. The need for scientifically based data to make informed decisions is vital if policymaking is to be credible. Likewise, policymaking is difficult when trying to support the common good because policy must placate many vested interests. Here is where education and training are vital for disseminating knowledge. For example, how does one figure out the best approach to implement vaccinations when people are unsure whether vaccines are safe?

Looking back over time, a general international consensus eventually emerged on social distancing, face masks, and hospitalization policies. It is hoped that what was learned and implemented during the COVID-19 pandemic will be useful in other crises. Finally, it should be remembered that public policymaking must not make things worse overall when solving problems.

Diversity in workforce behavior

COVID-19 altered the way business is conducted under lockdown as the economy ground to a halt and the economic pain of the pandemic grew. Everything from financing, to supply chains, to hiring and retaining employees comes into play when social isolation and work from home (WFH) becomes the norm. In the workplace, shutdowns force people to stay away from offices, causing a rethink of how to design office space. Staying away from the office causes many to enjoy the flexibility of work and family, as long as company productivity does not suffer. It should be mentioned that after three years, many workers have become tired of WFH isolation and wanted to return to the office. Thus, the jury is still out on which form of work participation is better. Recent return to office requests has had mixed results.

To achieve a dynamic workforce, Part IV suggests ways the business world might strive to increase diversity in the workplace, to ensure productive capacity in the economy. This process operates differently across generations as they enter the workforce with varying expectations, as described in the chapter by Bina Patel from the United States. Business organizations have to cater to employee needs in WFH, flexible attitudes, and childcare if worker wellness is to be maintained.

Six chapters are found under this theme. One chapter is about internships, freelance, and self-employment among entrepreneurs by S.R. Westvik from Norway. Bina Patel's chapter looks at generational shifts in work values and behaviors. There are two chapters on self-coaching and appreciative dialogue: Vikram Kapoor from the United States talks about self-coaching for pandemic survivors, while Keith Storace from Australia shares appreciative dialogue techniques. Marie-Louise Fehun Aren from Nigeria discusses the need for financial literacy to guide household savings among workers. Finally, Victoria Marquez-Mees's chapter relates to how management style efforts draw out collaborative teamwork at her workplace in a multilateral development bank.

Some have suggested that the youngest generation of workers is averse to 9–5 or a 5-day work week. They also seem to prefer WFH and collaboration versus individual approaches to work. If true, this may be based on not having yet a career profile; hence, they are willing to travel and not settle down. Others have suggested that the youth fear a future with limited options. Nevertheless, good news suggests that with demographic effects on early retirement, upward occupational mobility will increase for younger workers. Business leaders need to work toward transparency, collaboration and teamwork, access to fairness, emotional wellness, and work-family challenges including childcare, and in general opportunities for career growth that match business goals.

Technology and culture

In Part V, the cultural sector with help from technology was shown to keep creative ideas in the arts and sciences, flowing to provide purpose and dynamism in society. Virtual learning over the internet has to adapt worldwide to link up ideas cross-culturally for progress to occur. Just as education opportunity provides social mobility, the digital world of technology is transforming social media on the internet to allow more people to be creative and distribute content widely. At the same time, there is need to carefully monitor the use of personal information in big data operations to protect individuals.

Seven chapters found under this theme included one by Arthur Luna of the Philippines, questioning our overreliance on technology; a chapter by Li Mengxuan from China, looking at educational privacy issues and big data; a chapter by Lori Udall from the United States, looking at how artist creativity was stifled by COVID-19; a chapter by Jack Whitehead from the United Kingdom, looking into how the internet allows educational theory research to flow between East and West; a chapter by Tshering Cigay Dorji from Bhutan, looking at how technology supports gross national happiness in Bhutan; a chapter by Suresh Nanwani from Singapore, looking into personal growth and human connections; and a chapter by Alfred Anduze from Puerto Rico, examining how technology allows individual and community values to adapt in times of crisis.

Reading these chapters give the impression that technological innovation affects the way individuals interact with values and behaviors that keep society in balance. Each theme supports the function of each sector and is an important process that gets affected by external factors such as wars and pandemics. These chapters provide slice-of-life insights into how these themes operate daily to enforce sector goals.

Technological innovation in hardware and software delivery applied to the internet increased dramatically under COVID-19 lockdown, as gaming apps proliferated along with e-commerce and learning apps. The isolation of individuals at home forced virtuality as opposed to reality. Many groups were influenced by technology platforms differently. Teenagers found solace in gaming, most sought entertainment and e-commerce outlets, while others fed their fear and mistrust of authority through blogs. The use of avatars and gaming run on artificial intelligence algorithms altered the way people communicate using the latest ICT inventions. Young people became more tied to their smart phones because of the COVID-19 isolation. Social media, Zoom meetings, and e-entrepreneurial efforts changed ways on how people connected. In a way, COVID-19 accelerated technological innovation and changed the way we communicate and rely on artificial intelligence (Schemmer & Backes, 2015).

The COVID-19 crisis altered technological innovation to both speed up and adapt new ideas in the virtual world. Placing greater importance on machine learning in all facets of social activity will change the way society develops. This likely will become more pronounced as the metaverse of everything eventually alters the way people connect to each other. In a sense, the psychosocial theory of Erik Erikson discussed in Chapter 6[3] suggests both individuals and nations can lose their identity when social systems break down (Erikson & Erikson, 1998). COVID-19 forced isolation on individuals and institutions. This is not natural for humanity. Isolation can affect young children, adolescents, and young adults severely when building trust, initiative, and personal identity in the absence of school-going. Affected institutions within society include schools, hospitals, social media platforms, government services, large and small business enterprises, and the telecoms that support each.

Contributors' views on society and COVID-19

A description of how the survey of essay contributors felt about COVID-19 was reported in Chapter 2. While the survey of 27 contributors represents the lower limit of sample size in survey research, its advantage is the speed in cross-tabulating manually the interesting patterns of perceived notions among the participants at little cost. Our survey consisted of instructions to fill in a matrix found in Table 2.3 of key social activities processes, issues, and experiences of daily life influenced by COVID-19 drawn from the essays. Sixteen activities identified in the essays were selected and placed in a column of the matrix, including human connectivity, work-family stress, emotional IQ, and self-coaching. Respondents' views were sought on how important each one was in the COVID-19 pandemic. Four other columns were added, each representing the four themes of the book: home and schooling; public policy; workforce arrangements; and technology impact on social media.

Respondents were asked to assess each activity for each theme according to their own impressions of COVID-19 impact. Each activity could be rated across themes. Ratings on impact effects were organized as very important (5), important (4), moderate (3), small (2), and nil (1). In this way, respondents could assess each activity based on how COVID-19 impacted it across the four themes. Results from the matrix would (i) show an average impact for each activity, which would indicate their importance based on scoring across rows; and (ii) indicate which theme was most or least affected by COVID-19 based on averaging scores for each activity across each sector. In the case of big data issues, it is clear computer technology helped manage virtual education, crisis management of information flow, WFH arrangements, and provision of access to the internet and social media.

Based on findings gathered through the survey carried out in November 2022, along with media reports, it is possible to speculate on how common human activities will fare after the COVID-19 pandemic subsides. Taken together, these subjective impressions offer an overall assessment of just how much the pandemic changed society in subtle ways. Readers are encouraged to make their own subjective assessments for each item found in Table 2.3.

Drawing on the survey results from the contributors, two salient findings were revealed in Table 34.1. The first was that the contributors chose public health, personal connectivity, work-family, and big data as the most affected traits, while aging, cultural identity, and self-coaching were singled out as the least impacted by COVID-19. Social equality and research and development activities were almost evenly divided to have neutral impact.

It is not surprising that public health (along with personal health, to a lesser extent), personal connectivity, and work-family were rated highly. It was a bit surprising that big data collection was an area where the pandemic was felt to have greatly impacted society. The COVID-19 impact on cultural identity, aging, and self-coaching were collectively deemed less important by the group. Surprisingly, the respondents already seemed to be sufficiently self-assured in

TABLE 34.1 Number of contributors rating COVID-19 on themes

Twenty-seven contributors	*Educational opportunity*	*Policy-making management*	*Workforce behavior*	*Technology and culture*
	Number of ratings			
Number of top one and two (27 × 2 = 54) rating scores of themes by contributors	17	10	10	17

their cultural identity and need for self-coaching by not rating these activities as strongly impacted by the pandemic. The finding that many of the respondents (the majority is below age 65) did not consider aging as a core activity affected by COVID-19 suggests they may not have been aware of the severity of deaths and drama facing the elderly that comprised 65–70 percent of all COVID-19 deaths.

The second finding was that while contributors were evenly divided in writing chapters for each theme, they collectively agreed that COVID-19 greatly impacted themes one and four based on top two theme choices while themes two and three appeared least affected by COVID-19 when all 16 activities were averaged across each theme. Together core societal activities and processes were assessed across each theme to arrive at the impact of COVID-19 at the end of 2022.

Piecing together some patterns in the survey, we found slight differences among the contributors' perception of where COVID-19 affected society the most. Comparisons were made across East (14 respondents outside the OECD classification) versus West (13 respondents from North America, Europe, Australia, and New Zealand) regions of the world. Patterns revealed that in the East, personal connections and data were most often thought to be impacted by the pandemic, while the West focused on work-family stress and equality issues. The East strongly associated themes 1 and 4 with the pandemic, while the West weakly associated themes 1 and 3 with it (see Table 2.3 for characteristics and themes surveyed). In general, similarities surveyed on traits were not significant across gender and age, suggesting wide agreement by all 27 respondents and highlighting the impact of the pandemic on societal activities.

The book's contributors suggest some effort has been made in changing the way institutions accommodate public health emergencies, including social benefits, information dissemination, and resource allocation. WHO has pressed each nation to review and revise their protocols: (i) schools and hospitals have learned to prepare systems based on digital platforms; (ii) government has learned that well-publicized policy allays public fears; (iii) businesses learned they must modernize work environments to support employee mobility as well as be profitable; (iv) and culture has to employ ICT platforms more creatively to spread knowledge and values important to society. Time will tell if changes in personal connectivity, social mobility, emotional well-being, national resilience, and work-family balance will improve and thrive in the post-pandemic world of accelerating knowledge and technology linked to education.

Sector adaptability to COVID-19

The themes discussed in the book represent core processes at work in society that promote overall stability by reinforcing one another. The overall process is seen in Table 34.2.

TABLE 34.2 Sector-on-sector support to overcome COVID-19

Theme impact on each sector	Sector			
	Society	Polity	Economy	Culture
Theme 1: educational opportunity and social mobility	**mobility**	legitimacy	enterprise	purpose
Theme 2: public policy and risk management	participation	**governance**	regulation	experience
Theme 3: diversity in workforce behavior	dynamism	stability	**complexity**	expression
Theme 4: technology and culture	arts, sciences	goals	knowledge	**creativity**

- The social sector links to other sectors by providing **mobility** to the population. This provides the polity with legitimacy, the economy with enterprise, and the culture with purpose.
- The political sector links to other sectors by offering participation to the population. This in turn generates **governance** in the polity, regulatory safety in the economy, and freedom of expression in culture.
- The economic sector links to other sectors through **complexity** by allowing wealth and its accompanying lifestyle. This also provides political stability and creative expression in culture.
- The cultural sector links to other sectors by encouraging **creativity**. This allows goal direction in the polity, enterprise in the economy, and meaning to culture.

These sectors often work in tandem to reinforce one another when the public good is in jeopardy. In turbulent times resulting from the pandemic, these sectors can act as counterweights to crises as the social structure learns to accommodate existential threats.

Two Key COVID-19 takeaways requiring global attention

Educational opportunity and social mobility

The opportunity structures found across all sectors and themes from schools to workplace and cultural interaction on the internet allow people the mobility to choose and fulfill their ambitions in life. They are key to physical and emotional health. Under COVID-19, education and learning have been turned upside down (Breslin, 2023). In addition, health systems serve as early warning measures to detect illness for entire populations (Bismark et al., 2022).

Society should provide inexpensive and wide coverage educational platforms to support blended learning as combinations of educational technologies for both cognitive and emotional skill development. Opportunities for paid work, occupational choice, and income level should recalibrate business practices when balancing home-office integration that provides wider living styles offering freedom of choice (Kruse et al., n.d.).

At the same time, youth everywhere experienced severe disruption in their efforts to prepare for adulthood. In the process, many fell into depression, suicide, and loss of hope in their future prospects. Under lockdown that kept many teens at home, some became isolated from face-to-face contact with friends (Knight & Bleckner, 2023). Even worse, they were left with only partial online learning to complete their studies for over two years. This led to loss of learning as measured by tests. Clearly, isolation prevented socialization experiences needed for well-rounded personalities.

Finally, cultural mobility implies fluidity in combining various art forms that emote philosophical points of view. When ICT platforms and software spread cultural values across online social media, gaming, movies, and intellectual debate, they contribute to a virtual life in a blooming metaverse. New technology provides values and behaviors, social interaction, and differential experiences to create a virtual identity for the young to reinforce their ambitions and goals. The trend is to move from schools to internet to human experience in all areas of life found in the metaverse. Differences across many segments of society have increased during the pandemic, requiring social policy on COVID-19 across sectors and institutions to be carefully tuned so that skill acquisition and social mobility may exist everywhere for everyone.

Technology-driven creativity

The four themes interact with each other to support world development. Vaccine technology and medical responses, including telesurgery, research and development (R&D), genome sequencing, big data protocols, pharmaceutical marketing, vaccine production, telemedicine, hospital protocols, medical costing, publicizing, all support public health. These activities require high levels of education in the sciences, humanities, management, public policymaking, collaborative support, and group leadership to keep society running effectively under COVID-19 conditions (Loxley, 2019).

The culture that defines the social order has to maintain social cohesiveness and well-being during the pandemic to prevent isolation from causing depression and suicide. Social action requires big data for tracking and tracing infection rates, telemedicine, and human progress. In times of pandemics, the common good requires everyone to cooperate and not put others at risk. The best way to get consensus is to set agreed-upon rules to follow in public places where many people congregate and then exclude the few who, for whatever reason, cannot or will not comply.

Innovative ICT can help construct personal identity on the internet. In education, the opportunity to become someone successful happens when skills allow dreams to be realized digitally through lifelong learning. Likewise, government can pursue reliable data-dependent policy for the common good and disseminate findings for public debate. Policymaking is often the preserve of government in fighting pandemics. Yet governments were caught by surprise and did not know how to react quickly to COVID-19. Policies had to be put in place on a trial basis as information on the virus became available and transmitted to the public.

On the economic front, businesses use digital big data to improve efficiency for both profit and societal well-being, including the creation of employee-friendly workplaces. Previously, many business models were concerned with profits alone. New business models require looking at the well-being of employees in order to attract and keep talents. Improved human development creates valuable social and cultural experiences that contribute to the greater good of nations. ICT platforms may have great potential to push everyone into augmented reality that helps them become wiser and more self-confident.

COVID-19 changed the way society looks at the future of ICT, social mobility, and science research affecting the human condition. ICT has been both a blessing and a curse to many who survived the pandemic. The internet world provided news and entertainment but also unchecked time spent on social media with its spread of fake news and scams on the unsuspecting public.

For those learning online, opportunities increased to access learning, but after a while, motivation to study declined. Alternatively, socializing on platforms allowed for gaming and meeting online for those looking to make friends outside the traditional face-to-face meetings in school, work, and leisure. The downside of social media platforms was that scammers preyed on the innocent, and on the vulnerable in search of social relationships.

Core reforms needed in the coming years

Macroeconomic impacts of the pandemic have yet to take full effect as debt causes inflation and loss of productivity affects labor markets, schooling, and social media usage. Given the patterns discussed in the book, the pandemic drives social reforms in how to reorganize basic societal functions.

- Innovative learning systems are needed to disseminate knowledge to stay ahead of the global learning curve. Technological advances in platforms and software inventions are accelerating. Therefore, education and training must increase inputs and outputs both quantitatively and qualitatively. This effort is fundamental for societies moving into the digital age.

- Public policy needs to include stakeholder considerations in decision-making while finding ways to fairly distribute funding and delivery of public services. It also needs to have new modes of communication and networking based on big data that are inclusive and widely available. As importantly, public policy must also stress common goals that embrace the entire society. This is best accomplished by ensuring the nation has a well thought-out, practical, and publicly debated and agreed upon masterplan to move the nation forward.
- Workforce dynamics have changed in the new economy. Employer-employee relations are headed in new directions across the labor market, which include restructuring the workplace, work schedules, rewards and incentives, human resource development for workers in the gig economy, and frontline face-to-face occupations.
- A creative culture is essential if society is to remain dynamic and forward looking. With accelerating information technology in the digital economy, ICT allows both inclusive public goals and exclusive personal ambition to thrive side by side. This happens more and more through the internet as citizens gain entrance to the digital world of collaborative experiences. How will society build the metaverse from the internet to include a virtual world of activities with options for everyone? COVID-19 tells us to start now and include all. The areas that COVID-19 forced action during the lockdowns will carry forward with a life of their own and continue to alter key social activities of everyday life.

Conclusion

The disruption of COVID-19 pulled back the curtain of daily life. The pandemic made people more aware of their humanity that revolves around family, career, marriage, social connections, and wellness. COVID-19 raised expectations and made people aware of their mortality while challenging individuals to take power and seize the day.

The moral of the COVID-19 experience is not to waste life caught in a trap of mundane activities that lack purpose, which keeps people feeling unappreciated. Social isolation forced people into their homes and limited modern life. This made people feel a decline in their efforts to satisfy the hierarchy of social needs best exemplified by Abraham Maslow. The pandemic made people ask, "Stop the world, I want to get off" (which refers a musical of the same name by Leslie Bricusse and Anthony Newley).

However, this is not possible. Everyone must learn to live in reality that builds social networks that generate well-being, social interaction, and opportunities to lead useful lives. COVID-19 teaches us to (i) encourage the government to serve the public well; (ii) help educators design virtual education; (iii) support businesses to meet economic needs while securing opportunities for employee motivation for self-improvement; and (iv) monitor ICT to use artificial intelligence in ways that do not harm the individual or society. We all have learned a lot from COVID-19, and will continue to learn, to improve ourselves and thrive.

Notes

1 See Chapter 21, "How a Gen X became a Gen Z at heart" by Bina Patel, Chapter 22, "Unlocking from lockdown: reframing the future through appreciative dialogue" by Keith Storace, and Chapter 25, "Self-coaching for pandemic survivors" by Vikram Kapoor.
2 See Chapter 20, "Entering the workforce in the COVID-19 era" by S.R. Westvik.
3 See Chapter 6, "Tails we go, heads we stay" by Ariel Segal.

Reference list

Bismark, M., Willis, K., Lewis, S., & Smallwood, N. (2022). *Experiences of health workers in the COVID-19 pandemic: In their own words*. Routledge.
Breslin, T. (2023). Schooling during lockdown: Experiences, legacies, and implications. In J. M. Ryan (Ed.), *Pandemic pedagogies: Teaching and learning during the COVID-19 pandemic*. Routledge.
Charumilind, S., Craven, M., Lamb, J., Sabow, A., Singhal, S., & Wilson, M. (2022, July 28). *When will the COVID-19 pandemic end?* McKinsey & Company. www.mckinsey.com/industries/healthcare/our-insights/when-will-the-covid-19-pandemic-end
Collins, R., & Sanderson, S. (2010). *Conflict sociology: A sociological classic updated*. Routledge.
Erikson, E., & Erikson, J. (1998). *The life cycle completed*. W. W. Norton and Co.
Haas, J. (2021). *COVID-19 and psychology: People and society in times of the pandemic*. Springer.
Knight, K., & Bleckner, J. (2023). No magic bullets: Lessons from the COVID-19 pandemic for the future of health and human rights. In J. M. Ryan (Ed.), *COVID-19: Individual rights and community responsibilities*. Routledge.
Kruse, K. M., Leitch, J., Cooper, J., & Franken, A. (n.d.). *20 best new social policy books to read in 2023*. Book Authority. https://bookauthority.org/books/new-social-policy-books
Loxley, W. (2019). Liberal democracy trumps populism through education. In V. Jakupec & B. Meier (Eds.), *Abhandlungen der Leibniz-Soziatat der Wissenschaften in Die Auswirkungen des Richtspopulismus auf die Entwicklung des Bildungswesens* (pp. 115–140). Leibniz Society of Sciences.
Lupton, D., & Willis, K. (2021). *The COVID-19 crisis: Social perspectives*. Routledge.
Maslow, A. (2011). *Toward a psychology of being*. Amazon Books.
Parsons, T. (1977). *The evolution of societies*. Prentice Hall.
Parsons, T., & Smelser, N. (1956). *Economy and society*. Routledge.
Ressette-Crake, F., & Buckwalter, E. (Eds.). (2022). *COVID-19, communication and culture: Beyond the global workforce*. Routledge.
Rutherford, A. (2019). *The systems thinker*. Verband Deutscher Zeitschriftenerleger (VDZ).

Ryan, J. M. (Ed.). (2018). *Core concepts in sociology.* Wiley-Blackwell.
Schemmer, E., & Backes, L. (2015). *Learning in the metaverse.* IGI Global.
Wallerstein, I. M. (2004). *World systems analysis: An introduction.* Duke University Press.
Ying, L. L. (2022, November 11). World needs to start preparing for next pandemic now: Top WHO official. *The Straits Times.* www.straitstimes.com/singapore/world-needs-to-start-preparing-for-next-pandemic-now-top-who-official

INDEX

Note: Page numbers in *italics* indicate a figure and page numbers in **bold** indicate a table on the corresponding page.

Africa: Ebola outbreak, in West 8–9; financial illiteracy in Nigeria 209; possibility of new COVID-19 strains in 17; virtual hearings in 129–130; *see also* online tertiary education, Africa
Ahmed, Leyla 122, 286
Anduze, Alfred 256, 288
anxiety 38, 47, 49, 93, 159, 189, 195–196; in Australia 157; in baby boomers 260–261; in generation X 260; in generation Z 259; in millennials (generation Y)
appreciative dialogue (ApDi) 189; core principles 191; demographic overview and assessment tools 194–195; four interactional stages 193–194; framework 191, *192*; psychological structure 192–193; sequence *194*; therapy program 190
appreciative inquiry (AI) 190, 277–278
Ardern, Jacinda 109
Aren, Marie-Louise Fehun 208, 288
Argentina, infections in 35, 227
artificial intelligence 20, 99, 170, 222, 234, 286, 289
Asia Pacific Economic Cooperation (APEC) meetings 112–113

asynchronous learning 135, 237–238, 239
Australia 11, 35, 157, 286; governmental policy decisions and the law 158–163
Avian Flu 9

baby boomers 180, 182, 186; COVID-19 in Puerto Rico among 260–261
Bahamas, COVID-19 pandemic 46–53
Bayanihan (Republic Act 11469) 142–143
Beall, Michelle 252, *253*
Belgium 64, 285
Bethel, Owen 42, 44, 286
Bhutan 264; compassion in action 267; COVID-19 in 264–265; *Desuung* Skilling Program (DSP) 268–269; *Desuups* 267–268; Gross National Happiness (GNH) in 264; livelihood support 266; national preparedness and response plan 265–266
big data 41, 106, 151, 232, 234, 286, 287, 290, 293
Bird Flu pandemics 9
black swan event 3–4, 7, 105
Boudon, Raymond 57
Brazil, COVID-19 pandemic in 33, 115, 145, 118–121, 189, 186; Brazilian

institutions 119–120; infections in 35; pandemic and the world 115–116; pandemic deaths 35; role of multilateral institutions 116–117; structural disorientation 119; uncertainty 115–116

Caiado, José Guilherme Moreno 115, 286
Canvas 238
CFR (case fatality ratio) 259
Childhood and society (Erikson, Erik) 66–67
children 75, 257; challenges faced by 13, 16, 22; and parents in lockdown 158–159;
China, digital technology for COVID-19 policy in 149–150; digital technology in 154; grading system 153–154; Health QR Code 150–151; public management technology 155–156; Shanghai 152; Shenzhen 151–152; Travel Card 152–153; Venue QR Code 154–155; Xining 152
China online education: administrators 240; asynchronous learning 237–238, 239; collaborative learning 238, 239; engagement 240; online education and 237; parents 240; privacy issues in 237–240; relevant companies 240; schools 240; students 240; synchronous learning 237, 238–239
civil conscription, constitutional guarantee against 164–165
climate change 8, 52; post-pandemic 27–28
coaching: mental well-being and 284; self- **42**, 220, 221, 284, 287, 289, 290–291; space 220; witness 128
Cohut, Maria 203
collaborative learning 238–239
collective responsibility, individual liberty versus 63–64
compassion **24**, 220, 222, 262, 276, 277, 278; in action 267; and clarity of purpose 265–266; compassionate reason(s) 159, 160–161; lack of 257; self- 220, 222
conspiracy theories 46
constitutionality 122; technology deployment in judicial system 126–127
Coon, Rich 252, *253*

Cooperrider, D.L. 190
Coursera 237–238
COVID-19 pandemic 3–5, 44; adjustments in 2023 18–19; beginning of 5–7, 45–46; capacity for adversity 19–24; consequences on global development 28–29; cultural sector and well-being 26–27; economic sector and work 25; first year (2020) 10–15; global setting in 2019 7–10; political sector and public policy 24–35; post-pandemic relationship altering themes 27–28; pre-cursor to 44–45; second year (2021) 15–18; social sector and 24
creative culture 295
critical reflection in educational practice 246–248
cultural sector *26*, 26–27, 227–228; key issues arising from pandemic **24**; structural elements and **21**
culture: and entertainment 49–50; technology and 288–289; *see also* music; musicians

data privacy 161–162, 240
demographic shifts, post-pandemic 27
Dengvaxia scandal 145–146
depression 6, 23, 26, 46, 49, 58, 92, 189; in baby boomers 260–261; in generation X 260; in generation Z 259; in millennials (generation Y)
digital diplomacy 108, 112
digital technology 20, 87, 149, 243–249; for implementing COVID-19 policy in China 149–156; virtual platforms and 243–244; and Western critical theory traditions 244–245
digitalization, in India 83–84
Dimon, Jamie 204
diversity in workforce behavior 287–288
Doring, B. 190
Dorji, Tshering Cigay 264, 288
Draper, Simon 113

Ebola 8–9
e-commerce 170, 213, 289
economic diversification 52
economic sector 25, 169–171
educational theory research, COVID-19 in 243–249; critical reflection in educational practice 246–248; digital technology 243–249; educational

responsibility and equity 245–246; "educational", meaning of 245; human flourishing, values of 249; virtual platforms 243–244; Western critical theory traditions 244–245
e-internet information technology (IT), post-pandemic alter in 27–28
electronic case management system 124–125; demerits of 126; merits of 125–126; virtual court proceedings 126, 127
emotion, etiology of 256–257
emotional and physical isolation, in Latino community 256–263
employment *see* workforce, entering
entertainment, lockdown impact on 10–11
entrepreneurs 175–177
environmental protection, post-pandemic 27–28
Epps, Tracey 108, 286
Erikson, Erik 66–67, 71; stages of life 66–67
e-shopping 25
essays on COVID-19: aims and designs 32–33; essayists 37–39; international settings 33–34; worldwide deaths and infections *34*, 34–36; study structure and organization 36–37; survey responses and contributors 41–43, *42*; themes 39–41, *40*
Europe 33–34; infections in 35; pandemic deaths 35

family: life under COVID-19, in Peru 75–76
fear of contagion, in Peru 73
financial literacy 170, 208–217; and connection to educational curriculum 209–210; during COVID-19 210; financial losses from investment fraud 212; household debt levels 211–212; personal debt levels 211–212; personal finance components 214–217; personal financial literacy paradigm 212–214
financial losses from investment fraud 212
food, access in Peru 76
Forster, E. M. 275
France 65
Francis, T. 184
free trade agreements (FTA) 113

generation X (Gen X) 180, 182, 260; COVID-19 in Puerto Rico among 260
generation Y (Gen Y) 181–182; COVID-19 in Puerto Rico among 260; to gen Z 179–186
generation Z (Gen Z) 58, 183–184; COVID-19 in Puerto Rico among 259
Ghebreyesus, Tedros 35
gig economy 170, 295
global comparisons, with Philippines 145
global development: COVID-19 consequences on 28; prior to COVID-19 in early 2020 5–7
global economy 172, 265
globalization 7–8
governmental policy decisions, Australia: children and parents 158–159 court, challenges in 163–165; confusing government orders 157; data privacy 162–163; news and social media 158; stay-at-home orders 159–161; personal autonomy, rights and responsibilities 162; shortcomings of 162–163
grading system 153–154
Great Britain 45
great resignation 190
gross domestic product (GDP) 216
Guha, Shouvik Kumar 132, 286
guided meditations 222

Haggard, Merle 253
Happier Hour (Holmes, C.) 14
Hattie, John 58
Health QR Code, in China 150–151, *150*
healthcare system 17, 22, 34–35, 47–48, 52, 232, 292; collapsing 119; in Israel 68–69; public policy realigning 106–107
health-tracking systems 33
higher education: ApDi therapy for, students 191; in Inda 82–85; in London 62–63; in Peru 73, 76–77; *see also* virtual education
Hindu religion 277
Hoefel, F. 184
Holmes, Cassie 14
hope, after COVID-19 137
hotels, lockdown impact on 10–11, 34, 36
household debt levels 211–212

household products, access in Peru 76
human flourishing, values of 249

Ikigai 244–245, 276–277
income 214–215; protection 215
India 15, 17, 63, 285; digitalization 83–84; higher education in India 82–85; infections in 35; legal education in 132–139; pandemic deaths 33–34, 35; pandemic and technology 82–83; primary and secondary education 78–82; remote tribal areas 82; right to education 79–81; state bureaucracy 81–82; student absenteeism 84
individual liberty versus collective responsibility 63–64
Indonesia 33; pandemic deaths 35
information technology and health organizations 230–231
innovative learning systems 294
International Labour Organization (ILO) 6, 105
International Monetary Fund (IMF) 6, 105, 117, 120
international trade, post-pandemic 27–28
Israel 67–68, 70, 71, 285; healthcare system 68–69; vaccination initiative 69
Italy 11, 65, 172, 219

Japan 65, 145
Jones, George 253
journaling 221, 222
Junkanoo (traditional festival of costume) 50, 53

Kapoor, Vikram 219, 287
Kenya, COVID-19 and legal practice: alternative dispute resolution (ADR) 129; arbitral proceedings 129–130; constitutionality of technology in the judicial system 126–127; electronic case management system 124–126; new era in justice delivery 130–131; practice during COVID 123–124; practice pre-COVID 122–123; virtual court proceedings 126, 127–129
knowledge, strengthening 98–99
Korea 33

Latino community: emotional and physical isolation in 256–263; in Puerto Rico 258–259
La Viña, Antonio G.M. 140, 286

learning management systems (LMSs) 89
least developed countries (LDCs) 87
legal system: in Philippines 140–147; in Kenya 126–129
legal education in India 132–139
Lesotho 88–89, 92, 93, 285
lessons learned 184–186, 283–284, 295–296; diversity in workforce behaviour 287–288; opportunity and mobility 285–286; public policy and crisis management 286–287; reforms needed 294–295; sector themes 284–285; technology and culture 288–289; *see also* takeaways, key
life-choice experiences, post-pandemic 27
life and work transformation 202–203; lack of social network 203; new normal 206–207; team building 205–206; working from home 203–205
lifelong learning 95–100; consequences of 2020–2022 pandemic 96–97: education in metaverse 99–100; post-pandemic world 97; reproducing societal knowledge among youth 97–98; strengthening knowledge among elderly 98–99
Li Mengxuan 236, 288
Li Xudong 149, 286
lockdown 18, 22–23, 25, 35, 109, 140–141, 173; duration of 188; learning under 173–174; and suspension of education in Africa 88; unlocking from 188
London 175, 203, 205; higher education in 62–63; lockdown measures 63; vaccination campaigns 63
loneliness 67, 70, 169, 258; power of hug in 257–258
long-term quarantine, in Peru 73
Loxley, William 3, 57, 95, 105, 169, 227, 283, 285
Lucas, Robert 8
Luna, Arthur 230, 288

Maggiolo, Mark 251–252, *252*
malaria 9
management, issues faced by: lack of social network 203; new normal 206–207; team building 205–206; working from home 203–205
management flexibility 170
manufacturing, lockdown impact on 10–11

Márquez-Mees, Victoria 201, 288
Massive online open course (MOOC) versus Coursera 237–238
Mazurek, Nicole 157, 286
meditations 222, 245
mental health issues 188–189
Metaverse, education and 18, 99–100, 170, 228
Mexico 35; infections in 35; loneliness in 258; pandemic deaths 35
Middle East 4, 8, 11, 234
millennials *see* generation Y (Gen Y)
Milton, John 46
mRNA vaccines 11–12, 17
multilateral institutions, role in Brazil 116–117
music 228–229, 251–255
musicians, Virgina, USA 251–255

Nanwani, Suresh 3, 32, 57, 105, 169, 227, 244, 272, 283, 288
nations: health of populations in 8; high-population 34; issues after COVID-19 **24**; less-developed 35; poor 8, 15; rich 8, 15
Netanyahu, Benjamin 69
new normal 146–147, 206–207
New Zealand 6, 11, 35; 2020 to 2022 109–110; Asia Pacific Economic Cooperation (APEC) meetings 112–113; COVID-19 response 108–109; digital diplomacy 112; fortress country mentality 114; Free trade agreements (FTA) 113; goods exports 111; gross domestic product (GDP) 110; impact of COVID-19 response 110–111; international business 111–112; reflections 113; state of national emergency 109
news and social media 158
Nigeria 122, 212
non-governmental organizations (NGOs) 45

office space 170
office work, lockdown impact on 10–11
"off-the-shelf" technology 233
online education *see* virtual education
online education technologies (EdTech) in China 236; administrators 240; asynchronous learning 237–238, 239; collaborative learning 238, 239; engagement 240; online education and 237; parents 240; and related privacy issues 237–240; relevant companies 240; schools 240; students 240; synchronous learning 237, 238–239
online tertiary education, Africa 88–94; challenges for institutions 88–89; challenges for lecturers 89; challenges for students 90–91; first-year students 91–92; lessons learned 93–94 postgraduate research students 92–93; practicum activities 92; students in practical disciplines 92; WhatsApp platform 89–90

Pacific nations 15
Padlet 238
Pal, Anwesha 78, 285
Pan American Health Organization (PAHO) 46, 47
pandemic survivors: guided meditations 222; journaling 221; leveraging strengths and superpowers 221–222; meditations 222; observing positive changes 222; positive priming 221; self-coaching for 219–223
Paradise Lost (Milton, John) 46
parents, challenges faced by 5, 13, 36, 158, 170, 240
Patel, Bina 179, 287
personal autonomy 162
personal debt levels 211–212
personal development through appreciative dialogue 188–189: appreciative dialogue (*see* appreciative dialogue (ApDi)); appreciative inquiry 190–191; great resignation 190; growth mindset 190; therapy 189–190;
personal development, in times of crisis 61–62; collective solitude 62; higher education in London 62–63; individual liberty vs. collective responsibility 63–64; lesson learned 64
personal experience, of a student 65–66; adapting to life online 69–70; Erikson's theory and; in Israel 68–69; returning to normal 71; in Spain 66; understanding 66–67
personal finance components *214*, 214–217
personal financial literacy paradigm 212–214
personal mobility 59

personal relationships 275–276
Peru 72–77; family life under COVID-19 75–76; fear of contagion 73; food, access to 76; household products, access to 76; long-term quarantine 73; pandemic death rate 73; pandemic infection rate 73; virtual education during COVID-19 72–77
Philippines, COVID-19 and legal system 140–142; Bayanihan One/Bayanihan Two 142–143; Dengvaxia scandal 145–146; failure to adapt 144–145; global comparisons 145; legal framework 144–145; "no vaccine, no ride" policy 143; new normal 146–147
polio 9, 32
political sector 24–25, 105–106; change and perspective 50–51; COVID-related central themes 26; inaction of 8; key issues arising from pandemic 24; power-sharing consensus in 21; structural elements and 21
positive priming 221
positive thinking 278
post-pandemic: aftershocks 132–139; consequences of 2020–2022 96–97; education into metaverse 99–100; knowledge, strengthening 98–99; lifelong learning as powerful force in 95; post-world 97; youth, societal knowledge passing to 97–98
post-pandemic alteration: in climate change 28; in demographic shifts 27; in e-internet information technology (IT) 28; in environmental protection 28; in international trade 28; in life-choice experiences 27; in life-long learning 28; in science research and innovation 27; in social mobility 27
poverty and consumerism 210–211
primary and secondary education in India 78–82
public management technology, in China 155–156
public sector policy 39, 41, 106, 295; and crisis management 286
Puerto Rico: baby boomers 260–261; COVID-19 across generations in 258–259; generation X 260; generation Z 259; millennials (generation Y) 260

quarantine 5, 6, 10, 13, 22, 25, 274

reflexive legislation 141
reforms, needed 294–295
remote learning *see* virtual education
remote work *see* work from home (WFH)
remote tribal areas, in India 82
Republic Act 11469 (Bayanihan) 142–143
resilience 196–197
restaurants, lockdown impact on 10–11, 25, 34
Right of Education Act 85n1
right to bodily integrity 163–164
right to education, in India 79–81
risk management 105, 106, 285, **292**
risk protection 215
Roberts, Jim 253
Romer, Paul 8
Russia 33; infections in 35; pandemic deaths 35

Saco, Victor 72, 285
SARS 72
SARS-CoV-2 66
savings 214, 215
sector adaptability to COVID-19 291–292, **292**
Segal, Ariel 65, 285
schools, lockdown impact on 10–11, 13, 21, 22
science research and innovation, post-pandemic alter in 8, 27–28
self-coaching 221–223, 284
self-realization 277–278
senior citizens 95; demographic shifts 8, 27
Shanghai 152
Shenzhen 151–152
shopping, lockdown impact on 10–11
silent generation *see* generation Z (Gen Z)
Singapore 6, 65, 172, 288
Singh, N. K. 283
social isolation 3, 4, 37
social mobility 27, 39, 41; post-pandemic alter in 27 six essays 59; 28
social network, lack of 203
social sector: COVID-related central themes 26; and educational opportunity 24; introduction to 57–59; key issues arising from

pandemic **24**; and people 21; six essays 59; structural elements and **21**
social structure adaptability to pandemic 283
socioeconomic mobility 96
sociology 19–20, 27
South Africa 11, 17
South Asia 8, 15, 17, 34
South Korea, infections in 35
Southeast Asia prospects 233–234
Spain 65; drawing on theory 66–67; government's response 66; media's response 66
sport and fitness, lockdown impact on 10–11
state bureaucracy, in India 81–82
"stay-at-home orders" 159–161
Storace, Keith 188, 287
stress 74, 75, 159, 295; and anxiety 47; and depression 92
student absenteeism, in India 84
student learning 6, 36
Summers, Lawrence 283
super-stressors 201–202
Susan Pinker 207
Swine Flu pandemics 9
Switzerland 67
synchronous learning 237
systems theory 19–20

takeaways, key: educational opportunity and social mobility 292–293; technology driven creativity 293–294;
Taulelei, Rachel 112
teaching and learning 96
team building 205–206
technological business improvements 170
technology in Bhutan 269–270
technology and culture 227–229
technology driven creativity 293–294
technology platforms 231, 234–235; data for 232–233; providers 231–232; Southeast Asia prospects 233–234
telemedicine 262–263
Tencent Meeting versus Zoom 237
thirteeners *see* generation X (Gen X)
time on activities, between 2019 and 2020 *14*
tourism and travel, lockdown impact on 10–11
tourism sector 48
transport, lockdown impact on 10–11

Travel Card, in China 152–153
Tsheten, T. 269
Tsietsi, Tsotang 87, 285

Udall, Lori 251, 288
United Kingdom 36, 62, 115, 189, 203, 212, 243, 258, 288
United Nations 6, 11
United Nations Economic Commission for Africa (UNECA) 88
United States 8, 18, 45, 115, 118, 179, 258; infections in 35; pandemic deaths 35; Swine Flu pandemic in 9
university education: in India 82; in Peru 74
upward mobility 58

vaccine 17; COVAX 117; COVID-19 Vaccine Committee 47; Dengvaxia scandal 145–146; funding 106; in Israel 69; in New Zealand 110; production types of 11–12; and right to bodily integrity 163–164
values and family 180–181
Venezuela 67
Venue QR Code, in China 154–155
virtual education: higher education in India 82–85; institutional challenges in Africa 88–94; legal education in India 135–138; in Peru 72–77; primary and secondary education in India 78–82; privacy issues in online education, China 237–240
virtual classes, student participation in 136
virtual court proceedings 126, 127
virtual education, during COVID-19 pandemic in Peru 72
virtual platforms 243–244

Western critical theory traditions 244–245
Westvik, S.R. 172, 287
Whitehead, Jack 243, 246, *247*, 288
work environment, diversity in 39, 41
work from home (WFH) 13, 22, 25, 170–171, 182, 203–205; accelerated by pandemic 220; affecting interactions 204–205; as career change prompt 183; diversity in workforce behaviour 287; earning more with 213; IT development for 230–231; in Kenya 124; returning to "next normal" 71

work-family balance 64, 170, 177
workforce behavior, diversity in 287–288
workforce dynamics changing 185–186
workforce, entering 172–174: COVID-19 impact on students' career paths 174–175; entrepreneurs 175–177; new world changes 177–178
World Bank 87, 117
world development, and COVID-19 6, 293
World Health Organization (WHO) 4, 11, 25, 33, 35, 46, 47, 105; statistics on COVID-19 pandemic 6
World Trade Organization (WTO) 105, 116, 111
world, after pandemic 97

worldwide deaths and infections, actual rate versus real rate 34–36
writers: approaching 39; characteristics 38
Wuhan, China 11, 149

Xining 152

yin and yang mentality 7
yoga 244–245
youth: challenges faced by 8, 16, 22–23; societal knowledge passing to 97–98
Ysewijn, Frédéric 61, 285

Zero-COVID Policy 149